From the visionary founder of the Self-Assembly Lab at MIT, a manifesto for the dawning age of active materials

THINGS IN LIFE TEND TO FALL APART. Cars break down. Buildings fall into disrepair. Personal items deteriorate. Yet today's researchers are exploiting newly understood properties of matter to program materials that physically sense, adapt, and fall together instead of apart. These materials open new directions for industrial innovation and challenge us to rethink the way we build and collaborate with our environment. *Things Fall Together* is a provocative guide to this emerging, often mind-bending reality, presenting a bold vision for harnessing the intelligence embedded in the material world.

Drawing on his pioneering work on self-assembly and programmable material technologies, Skylar Tibbits lays out the core, frequently counterintuitive ideas and strategies that animate this new approach to design and innovation. From furniture that builds itself to shoes printed flat that jump into shape to islands that grow themselves, he describes how matter can compute and exhibit behaviors that we typically associate with biological organisms, and challenges our fundamental assumptions about what physical materials can do and how we can interact with them. Intelligent products today often rely on electronics, batteries, and complicated mechanisms. Tibbits offers a different approach, showing how we can design simple and elegant material intelligence that may one day animate and improve itself—and along the way help us build a more sustainable future.

Compelling and beautifully designed, *Things Fall Together* provides an insider's perspective on the materials revolution that lies ahead, revealing the spectacular possibilities for designing active materials that can self-assemble, collaborate, and one day even evolve and design on their own.

"In Skylar Tibbits's ideal world, roads, buildings, and objects are tingling, made of active materials whose particles and units bind and unbind and recombine in mesmerizing harmony. There is little to no waste, an endless trove of new forms and solutions, and the ability to test and perfect along the way. I want to go there."

Paola Antonelli, senior curator of architecture and design and director of research and development, Museum of Modern Art

"In this book, Tibbits proposes a future where artificial intelligence is not an end in itself but an embodied feature of the products that we make. It is a future that is more humane precisely because of the shared tactility and materiality of stuff. There is no doubt in my mind that the future of materials science lies in the development of the types of animate matter described in this book."

Mark Miodownik, author of *Stuff Matters*

"*Things Fall Together* is a revolutionary book that helps us see into the future. Skylar Tibbits provides new design possibilities that rely on biological principles to activate materials into self-assembly. His pioneering approach is exactly what we need for Mars exploration and other space missions."

Dava Newman, Apollo Professor of Aeronautics and Astronautics at MIT and former NASA Deputy Administrator

Things Fall Together

A Guide to the New Materials Revolution

Skylar Tibbits

PRINCETON UNIVERSITY PRESS

PRINCETON AND OXFORD

Princeton University Press is committed to the protection of copyright and
the intellectual property our authors entrust to us. Copyright promotes
the progress and integrity of knowledge. Thank you for supporting free
speech and the global exchange of ideas by purchasing an authorized
edition of this book. If you wish to reproduce or distribute any part of it in
any form, please obtain permission.

Requests for permission to reproduce material from this work
should be sent to permissions@press.princeton.edu

Published by Princeton University Press
41 William Street, Princeton, New Jersey 08540
6 Oxford Street, Woodstock, Oxfordshire OX20 1TR

press.princeton.edu

All Rights Reserved

Library of Congress Cataloging-in-Publication Data
Names: Tibbits, Skylar, author.
Title: Things fall together : a guide to the new materials revolution /
    Skylar Tibbits.
Description: Princeton : Princeton University Press, [2021] | Includes
    bibliographical references and index.
Identifiers: LCCN 2020036342 (print) | LCCN 2020036343 (ebook) |
    ISBN 9780691170336 (hardcover) | ISBN 9780691189710 (ebook)
Subjects: LCSH: Programmable materials.
Classification: LCC TA403.6 .T53 2021 (print) | LCC TA403.6 (ebook) |
    DDC 620.1/1—dc23
LC record available at https://lccn.loc.gov/2020036342
LC ebook record available at https://lccn.loc.gov/2020036343

British Library Cataloging-in-Publication Data is available

Printed on acid-free paper. ∞

Printed in China

10 9 8 7 6 5 4 3 2 1

42

Contents

Acknowledgments

A SINCERE THANK YOU to those who have contributed to this book directly or indirectly, everyone who collaborated with us on research, and the many people who inspired and supported me throughout the process! This book would not have been possible without the amazing dedication from Princeton University Press, specifically Jessica Yao, Eric Henney, Chris Ferrante, and Madeleine Adams. Thanks for believing in the book, continuing to push me throughout, and helping turn ideas into reality. Thank you to Patsy Baudoin for your continual advice and guidance over the years on all of my books. Thank you to my MIT community, Terry Knight, Andreea O'Connell, Nicholas de Monchaux, Hashim Sarkis, Andrew Scott, Meejin Yoon, Ana Miljacki, Leila Kinney, Patrick Winston, Erik Demaine, and everyone in the Department of Architecture, the School of Architecture and Planning, the International Design Center, CAST, and many others. Thank you to everyone who contributed to the book: Tal Danino, Manu Prakash, Peng Yin, Fiorenzo Omenetto, Rob Wood, Jennifer Lewis, Radhika Nagpal, Fabio Gramazio, Matthias Kohler, Marcelo Coelho, Casey Reas, Ben Fry, Daniela Rus, and Suzanne Lee.

I'm forever grateful to our entire team of amazing researchers at the Self-Assembly Lab, including Bjorn Sparrman, Athina Papadopoulou, and my codirectors, Jared Laucks and Schendy Kernizan. Without our incredible team, the work would not be possible or anywhere close to as awesome! Thanks to our nonhuman material collaborators, because you are the true designers, keeping us on our toes and surprising us with realities we couldn't have foreseen. Thank you to our human collaborators: Christophe Guberan, Hassan Maniku, Sarah Dole, Doug Holmes, Art Olson, Marcelo Coelho, Neil Gershenfeld, Tom Claypool, Gihan Amarasiriwardena, Gramazio Kohler Research, Patrick Parrish Gallery, ICD, Ministry of Supply, AFFOA, Steelcase, AWTC, Ferrero, Airbus, Carbitex, Native, BMW, Stratasys, Autodesk, Google, Tencate, and many others. Last but not least, to my parents, D and J, and my family, V and Z and R: thank you for all of your support over the years and putting up with this neverending book project!

Things
Fall
Together

Programming Matter

IN THE EARLY 1700S, the English carpenter and clockmaker John Harrison solved one of the most vexing puzzles that sailors faced at the time: how to calculate longitude while at sea. This challenge was so important for navigation—and had been so confounding up to that point—that the British Parliament offered a substantial cash reward to anyone who could find a practical solution. As trade increased, and ships sailed around the world with increasing regularity, it was critical for the crew to understand where exactly their ship was along the earth's horizontal axis. Disrupted by the challenging conditions at sea, timekeeping and way-finding devices were inconsistent and unreliable. Consequently, navigation at the time was notoriously imprecise and shipwrecks were far too common as a result of ships losing their way.

While scientists and many others looked to astronomy, mathematics, or even magic in their quest to unlock an answer to the riddle, Harrison's solution was amazingly simple and elegant. From wood, metal, and other simple material components, he crafted a "sea clock" that could keep reliable track of the time in relation to a given reference location, which would allow sailors to calculate their position based on the difference from their local time. Earlier attempts at such clocks had been thwarted by the motion of the sea, changes in the environment, and accumulating errors in the mechanical clockwork. But Harrison's design, by accounting for the ways in which materials would expand and contract, enabled his mechanism to adapt naturally to even the most minor fluctuations in temperature, pressure, moisture, and physical movement. As a master craftsman, Harrison understood that the dynamic and adaptive properties of his materials were the keys to a sea clock that could keep perfect time for long intervals, no matter the weather, the conditions of the sea, or the movement of the device.[1]

His invention became known as the marine chronometer, and it revolutionized not only sea navigation but also the way we think about materials and their ability to adapt in intelligent ways. Harrison demonstrated how material properties could be exploited to solve notoriously challenging design and engineering problems. Since that time, similar material-based mechanisms have been applied to a number of novel devices that are abundant in our everyday lives. Thermostats, for example, take advantage of a bimetallic structure to regulate the temperature in our houses or maintain safe operating temperatures in an engine. Orthodontic devices are made from Nitinol, a nickel titanium alloy that can move teeth into precise locations based on a response to body temperature. Lifesaving medical devices like stents use similar bimetallic structures to morph from one shape into another. This behavior has been "preprogrammed" in the material through heating and molding it at high temperatures. When a stent is placed in the body, for example, it is collapsed to fit through small spaces, and then activated by body temperature, allowing it to morph into the memorized shape and open the vessel.

Yet this way of working with materials to craft elegant, simple, and transformative solutions is still largely contained to a few niche applications, and not widely used today. Since Harrison's time, we have moved from a society that produced goods with localized crafts-based knowledge—one in which products and environments were intimately and intrinsically linked with material properties—to a system of industrially standardized mass production. The Industrial Revolution effectively ignored the intimate material knowledge of previous generations. Instead of taking advantage of the inherent material properties within wood or metal, for example, factories started to create standardized components that attempted to limit the amount of heterogeneity and differentiation. We attempted to standardize the trades and create repeatable outputs that did not rely on a single person's skill set or knowledge in the craft—with some good reason: it was much more difficult to make a house out of logs and branches, or a stone wall out of geometrically unique elements, than it is to construct anything with repeatable components like bricks or two-by-fours. Similarly, at an environmental scale, humans shifted from an intimate relationship working with the earth and the natural forces of rain, sun, storms, tidal shifts, or sediment movement to a top-down, brute-force dictation through the use of machines. We could build anywhere, create land, dredge, redirect water flows, and artificially construct nearly any environment. Most of this standardization in manufacturing, construction, and land use was attempting to fight the dynamics of materials, minimizing their movement, and resisting the forces of the environment (gravity, temperature changes, moisture changes, vibration, natural disasters, and so on). The goal was to produce more, and to do it faster, cheaper, and better.

This alienation from materials has only been exacerbated in recent times by the rise of computing and the digital revolution. Digitalization and virtualization have tended to disconnect the average person from materiality and led us to believe that creating something "intelligent" means either a human being or a digital system with software/hardware that simulates human intelligence. But all of our own human and biological intelligence is ultimately built from simple materials, not computer chips or robotic components. We have lost touch with our appreciation for material intelligence.

I often think of Harrison and his marine chronometer and wonder: if society were challenged with the same problem today, would we come up with the same elegantly simple solution? Hundreds of years later, simple devices like this can encourage all of us to take a fresh look at the way we design with materials, even as new research and technologies have us poised to surpass traditional craft-based production methods. The emergence of digital fabrication technologies and the rapid advance of new research in synthetic biology, materials science, and other fields are making it possible not just to tap into, but also to create material properties in a new way, bringing the possibility of a new industrial revolution into view—a materials revolution.

In this book, I offer you a glimpse inside this emerging materials revolution, from my vantage point as founder and codirector of MIT's Self-Assembly Lab.[2] The Self-Assembly Lab is a group of architects, designers, artists, engineers, scientists, computer scientists, and many others who work on a variety of research topics from self-assembly to new material behaviors or new fabrication processes. Through this work, we explore applications in product design, manufacturing, construction, and large-scale environments. Sitting at the intersection of design, science, and engineering, we are an academic research lab that blends creativity with exploration, elegant design aesthetics with technical performance, and the design principles needed to make those ideas reality. At its core, our work is motivated by the conviction that smarter, higher-performing products and sustainable environments don't require complicated, expensive, device-centered solutions to achieve. Instead, we seek to use simple materials and their relationships with environmental forces to design and create a more active, adaptive, lifelike world around us.

In this work, we are part of a broader community of scientists, engineers, and designers across research and industry who are finding ways to design, create, and program physical materials that can do more than even Harrison could have dreamed. These materials can take in information, perform logical operations, sense, react, and much more. Unique behaviors often seen only in living natural systems—like the ability to correct errors, reconfigure, replicate, assemble themselves, grow, evolve, and so on—can now emerge in innate material objects. At the Self-Assembly Lab, for instance, we have explored phenomena where physical components assemble and self-organize to build structures from objects, furniture, electronic devices, and even land formations. By understanding and utilizing material capabilities, we can give simple materials and environments new functionality— going beyond mass production or even mass customization, into material programmability with behavioral intelligence built into our products.

As we will explore throughout this book, recent material advances are influencing various fields from robotics to apparel, furniture, medical devices, manufacturing, construction, and even coastal engineering. With novel material functions embedded within fibers, we are now creating clothing and textiles that can adapt to temperature or moisture fluctuations and keep you cool or dry on the fly. Furniture and products can transform in size, shape, or function and assemble themselves after being shipped flat. Novel medical devices are emerging that can be quickly multimaterial printed to be customized to the individual's body. When they are inserted, they adapt to the person's internal environment, expanding arteries or air passages without complex behaviors. At the largest of scales, a simple material like sand becomes a medium to promote the self-organization of new islands or coastlines by tapping into the energy of the ocean. These and many other material-driven performances are coming into reality where simple products are becoming more active and static things are becoming more lifelike and playful.

This kind of work ultimately requires a new way of collaborating with materials in our broader environment, new relationships with our products, a different mindset, and a fresh way of looking at the world. This book describes that new mindset through simple design principles that offer new ways to think about traditionally "static" mechanisms, products, and environments—as well as a different definition of what makes a product "smart." The world is crying out for highly intelligent, active, and "smart" products, yet far too often we see smart products that are expensive, complex, battery-powered devices that are prone to failure. The principles in this book point to a different path forward. My hope is that they will make you stop and think, and wonder why some "smart products" might not be quite that smart after all. The aim is to show how we can take advantage of these hidden possibilities inherent in our physical world—and uncover a new relationship with materials, tapping into their built-in intelligence.

What do we mean when we talk about programming materials, and how has this reality emerged? We can start with a general definition: to program something is to create a set of executable instructions that an intended medium can perform or process. This is, obviously, a *very* general definition of programming—I'm using *medium* instead of computer, because, as I will explain, *we can embed a program into any medium.* Any time we perform a set of instructions, we're executing a type of program. When we program materials, we're embedding such instructions into a *physical* material, such that the material can make logical decisions and can sense and respond to its environment.

Thus, we can define a **programmable material** as a physical material structure that is embedded with information and physical capabilities like logic, actuation, or sensing. A related term, **active matter**, is used throughout this book to describe the expanded field of researchers that are programming materials from the smallest to the largest of scales to create highly active structures that can self-assemble or physically transform.[3] I will both describe the ways to program a material and explore the applications of its active behavior. In essence, these emerging material systems are all based on the ability to take simple material components, activate them with energy, and then have them assemble, transform, and create new physical behaviors.

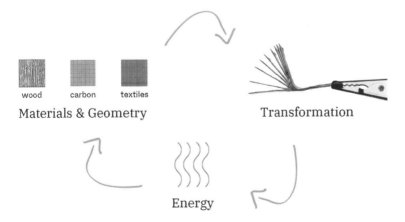

wood   carbon   textiles

Materials & Geometry   Transformation

Energy

A diagram showing the key ingredients for programmable materials: materials, geometry, and energy to create physical transformations. *Credit:* Self-Assembly Lab, MIT

The idea of matter that can be programmed is a fairly old concept, but our understanding and realization of the idea has changed. People have been dreaming of programming matter since at least the Star Trek Replicator, with a machine that could instantly create anything.[4] There are many early examples from science fiction that dream of infinitely small programmable material units that can be easily fabricated and set free to live, grow, and transform.[5] This dream has a long history of over-promising and underdelivering, however, most likely due to the lack of material and fabrication capabilities, until very recently.

Of course, from another perspective, there is a sense in which matter has always been "programmed." Everything around us is *programmed* to sense something or function based on built-in information. The most obvious examples come from living systems: just think of our DNA, which encodes the instructions to build a human, or how a plant grows toward sunlight. But our everyday life is replete with materials that transform in this manner, according to built-in information. In addition to complex living things, we can also see physical transformation in natural, yet nonbiological materials, or even synthetic materials that sense and respond to the ambient environment. For example, crystals that grow and morph, or chopped wood, which is no longer alive yet will still warp in response to changes in humidity, and plastics that expand or contract based on temperature. All of these materials are nonbiological, coming from both natural and synthetic systems, and all demonstrate lifelike, information-rich behavior.

Craftspeople, master builders, or anyone with an intimate, hands-on relationship to materials like John Harrison are the forerunners of today's "matter programmers," having long taken advantage of the inherent characteristics of materials. For example, craftspeople have used wood's inherent properties when making furniture or building joints, ship hulls, or whiskey barrels, forging tighter and stronger joints by changing the amount of moisture in the environment in which they were made. Metalsmiths often use the expansion and contraction of metal based on temperature to make precise and strong connections. Or engineers design a metal component for an engine to be able to operate uniformly with ever-changing environmental fluctuations. Textile manufacturers often use temperature and moisture to control the contraction of a garment to create finely tuned shapes and sizes.

Today, however, new digital fabrication technologies can produce at speed and scale while also customizing material properties, giving us greater capabilities than ever before. Computing, fabrication, and materials share deep and long-standing links. The Jacquard loom, invented in the 1800s and considered one of the earliest examples of computing, read punch cards as an analog program to create intricate and beautifully complex woven textiles.[6] More recently, not long after the modern computer was born with the invention of the transistor in 1947, scientists at MIT first linked a modern-day computer with a milling machine in 1952.[7] This paved the way for the first computer-aided design (CAD) tool in 1963 and today's computer numerically controlled (CNC) machining with CAD to CAM (computer-aided manufacturing) design workflows. This allowed computers to be programmed to run fabrication equipment that produced material parts.[8] Nearly every manufactured product today is made in some way with this sequence of technology—CAD to CAM to CNC—from electronic devices to cars, clothing, buildings, infrastructure, airplanes, and even children's toys. In the twenty-first century, we have achieved an ever-increasing level of sophistication with digital fabrication capabilities using laser cutters and water jets, 3D printers, milling machines, industrial robot arms, and many other technologies. These are acquainting more and more people with the properties of materials and machines, as well as eroding traditional boundaries between design and fabrication, materials and information.

This development in computing, fabrication, and materials research has led to the growing materials revolution and enabled programmable materials. Not only can we take advantage of the hidden abilities within materials to sense and transform, but we can customize the material with these rapidly advancing fabrication techniques. In the same way that we can alter the "instructions" coded into DNA using principles of synthetic biology and other technological advances like gene editing, we can now customize and produce complex compositions of many different materials, from scratch. We can go beyond the evolutionary mutations that have led to specific genes or material properties to now fabricate embedded material codes. For example, we can now produce synthetic wood that responds to moisture with customized grain patterns that would never be found naturally, complex metal components that adaptively tune engines, high-performance composites that morph for aerodynamics, and multimaterial printed structures for smart medical devices. All of these examples have both geometric complexity and a diversity of material properties, designed according to a set of instructions for tunable and adaptive performance.

This entire progression, from naturally evolved materials to synthetically designed and generated materials toward fully programmable materials, can be seen in the simple example of the continuing evolution of fashion and footwear. We can trace the lineage from traditional natural spun-cotton garments to programmably controlled textile production, and now synthetic fibers and high-performance textile products. More recently, the industry is turning the corner toward smarter garments that have sensors and actuators embedded within the textile to inform and act on our body's every move. These robotlike articles of "smart" clothing are quickly evolving from bulky garments with bulky devices to simple and smarter materials. The Self-Assembly Lab has worked with emerging companies, like Ministry of Supply, to develop highly active garments that can be made from material properties intelligently knit into intricate garments, functioning through materials, rather than complicated devices.[9] Garments can become porous and breathe when the person is hot, or get thicker to insulate them when they get cold. The garment can morph to the person's body shape and create the perfect fit, or change aesthetics for different occasions. Not only are we using novel materials, but we can now fabricate garments and other products in this new way, enabling active performance in everyday clothing.

One might assume that a *programmable* material would be more electronic or robotic, less human, static, and less active—just sitting there waiting to be programmed. But as I hope to show you, today's *digital* capabilities have actually reintroduced the human perspective and the craft of materials. Intimate knowledge of a material's properties brings surprise and intuition back into discovery and invention. Programming materials is more about opening our eyes and designing in collaboration with materials rather than forcing them into place.

The ideas within this book seek to illuminate the surprising, yet still mostly untapped, capabilities of materials through novel approaches to design and fabrication. We will uncover ways to seemingly reverse entropy, create simple material "robots," and program everyday physical objects or environments to come alive. We will challenge the conventional idea that things fall apart: objects can get better with time, and we can program materials to become more active, to adapt, and to evolve on their own. We will question why so many objects and environments are designed to be static and why human-made things typically don't have lifelike properties—for example, why they can't grow, transform, or repair themselves. Why does a "strong" structure usually mean it requires *more* material, *more* rigidity in its composition? Think of a plant or a tree, whose strength usually comes not from bulk or excess material, but from efficient distribution, flexibility, and the ability to adapt to different forces, to error correct, or regrow when needed. We will discuss the reasons why we've become so comfortable with the notion of what a robot or a computer looks like and how it behaves, yet why that is rapidly changing. In this way, we arrive at the new reality of active matter.

These ideas have taken shape after years of play, experimentation, collaboration, failure, and some happy accidents at the Self-Assembly Lab, yet they go far beyond our own work, crossing many academic disciplines and offering surprising applications in many different fields. This emerging field is based on blending rigorous science and engineering with creativity, play, and imagination. Progress requires not only the solution of technical problems but also the freedom to explore creatively and to take big risks, tackle big questions, and propose radical ideas. Accordingly, throughout the book we'll explore both concrete examples of technological advances being made today by talented designers, scientists, and engineers across different fields, as well as near-term thought experiments and possible futures. While this emerging field is rapidly growing and has shown promising advances, it is still early days. We are in just the beginning of this materials revolution, and much of the potential impact or applications have yet to be realized. At this exciting juncture, I am hoping to create purposeful visions for the future to help catalyze these advances, inspire applications and new collaborations, and energize the field of active matter.

Computing
Is Physical

WHEN WE LOOK AT A COMPUTER, we see a machine that translates the physical to the digital. Our busy keystrokes are transformed into digital information (0s and 1s), making its way through silicon, transistors, chips, and drives into computation and communication. This machine translates our conceptual and abstract thoughts into digital information through physical buttons and then patterns of 0s and 1s. It takes instructions, executes those instructions, performs logical operations, and magically transforms information from physical input to digital data.

Consider now moving from the digital to the physical again: we can connect our computer to a printer to produce physical paper documents or, nowadays, we may connect our computer to a 3D printer and fabricate physical objects. Or we could connect a computer to a motor, and send code to tell the motor to move something. An ever-growing world of software and hardware eases this translation from the digital to the physical. Platforms like Arduino, Makey Makey, and Little Bits translate computer code into the physical environment, allowing anyone to easily make interactive electronic devices.[1] Blurring the boundaries between physical/digital and human/machine has exploded in recent years due to emerging fields like human-computer interfaces (HCI), haptics, user experience and interface design (UX/UI), interaction design, physical computing, and many others.

Yet, early forms of computation involved no such boundaries. Computers were *only* physical—in other words, there was no translation process when it came to the actual computation. For centuries, computing and calculating meant using an abacus, the knots of a *quipu,* or a handful of pebbles. Likewise, some of the first "computers" were actually the brilliant women at NASA who played a vital role in making complex calculations to help land humans on the moon.[2]

Early computing machines—from the Jacquard loom and Ada Lovelace and Charles Babbage's Difference and Analytical Engines in the first half of the nineteenth century to Vannevar Bush's Differential Analyzer a century later—were built with gears, pulleys, belts, or vacuum tubes, each meant to translate and calculate. We consider these machines "analog" when we compare them to today's computers because they were noisy—not just in the literal sense of the word, but in terms of their material tolerance, faulty components, and mechanisms. The performance of these machines decreased as they got larger and more complex. The more gears and complex components, the more likely they were to break. The longer the machine ran—the more likely it was to break. As the length of time for a specific computation increased, the accumulated tolerance and errors increased because the small physical errors in a gear or pulley, for example, could cumulatively make things become way off hundreds or thousands of cycles later. These early computers were affected by the temperature or moisture of the room, literally expanding and contracting, causing unintentional changes in performance or introducing maintenance errors. Early computers were so very material that the first computer "bug" was said to have been found inside Harvard's Mark II, which was a literal moth trapped within the machine that caused an error.[3]

The concept of "digital" computation came into being around the 1940s, when the first notions of digital information and communication were laid out in one of the most influential theses of all time by Claude Shannon, showing how to communicate reliably with unreliable devices.[4] Shannon's work outlined principles for how to translate information from one place to another in a reliable fashion. These methods can be applied to physical, macroscale materials like mechanical computers, or translated to electronics in silicon today. Most relevant at the time, however, was how we communicated across noisy telephone lines in the 1940s. At the core of his theory was the idea that as information is sent from point A to point B, just as in a game of Whisper Down the Lane, errors can easily creep in. But if you create a system of error correction, like checks and balances, where you send multiple copies of the same information, and then cross-check the information on the other side, you can reconstruct the information and ensure it is perfectly accurate. This became the essence of being "digital" versus analog.

Even the first computers that were considered "digital," incorporating Shannon's principles of error correction and programmability, were still made mostly from mechanical parts with the addition of electrical switches. The computers used by Alan Turing and the code breakers at Bletchley Park, like the wartime Enigma-breaking machine, Colossus, and John Von Neumann's ENIAC (Electronic Numerical Integrator and Computer), were made with vacuum tubes and still relied on paper tape or manual cable switching as input.[5] These digital machines were slow, difficult to program, and physically enormous, eventually leading to their replacement with ever more modern forms of computers, moving from gears and mechanisms to silicon and transistors with the invention of the transistor at Bell Labs in 1947.[6] These smaller and faster transistor-based computers eventually led to desktop computers in the early 1980s. Our modern computers today maintain accuracy in computation and communication even as time passes or as the distance increases between the sender and the receiver, in part thanks to Shannon's work in the 1940s. However, as a result of the miniaturization and reliability of computer hardware, our idea of computing as inherently physical has simply disappeared.

Alongside the development of computers, the first CNC machine at MIT in 1952 paved the way for computers to run personalized fabrication equipment. This led to the development of CNC milling machines, laser cutters, water jets, 3D printers, and many other tools that have become easily accessible for individuals or factories.[7] Just as code became a new tool to design and create, code has enabled new forms of physical making to turn ideas into reality. In the past few decades, as the digital fabrication and maker movements have exploded, we are seeing more and more physical making in fab labs, tech shops, and maker fairs, where people have become acquainted with the properties of materials. This development led in part to the materials revolution that is upon us. With the growing interest in materials simultaneously coinciding with new biological and synthetic material discoveries, it is now possible to create digital and programmable material just like their computer-machine counterparts.

In recent years, a *digital material* has been defined as a set of material parts that can be precisely, yet reversibly, assembled and disassembled.[8] This can apply to DNA, Lego building blocks, press-fit assemblies, or many other discrete building blocks. This definition of digital materials doesn't fully address material properties directly, however. The evolutionary path that enabled material programmability certainly involves the building blocks, but it is even more ingrained in the material's properties. Programmable materials go beyond digital information in a material and are more than just the discrete assembly or disassembly of components. The extraordinary properties inherent in any physical material can contain information, sense a stimulus, and respond in sophisticated ways. This takes us beyond digital material to fully programmable materials.

Today, we tend to associate computing solely with a laptop or a phone, losing sight of the material properties that make computation possible. Perhaps it is because we cannot see the inner workings of today's digital computers: the components are so small, and the invisible electron-based processes on which they rely are immensely complex and hard for everyday users to grasp. Most of us don't really understand how it works; we take it for granted, almost as if it were magic. Or perhaps it is the sheer ease and speed of modern computation. We type a line of code or send a quick email and forget that very specific material properties—of silicon in the transistor, of solder in the circuit board—and intricate geometries within the circuit are enabling this simple computation. As Moore's law has increased the number and decreased the size of the components on a chip, a profound disconnect has grown between the physical computer that we type away at and the tiny components that enable computation under the surface.

It is easy to understand why we were once fooled into thinking that the future is digital and not physical, as our world becomes increasingly virtual with more computing in our pockets, daily video conferences, and virtual reality surrounding us.[9] Yet, as we all feel strangely less and less physically connected while being more and more digitally connected, we are realizing that the digital does not necessarily need to be less physical. The loss of physical connection and tangible realities from our devices has actually created a renewed need for the physical. If we look at the COVID-19 pandemic, for example, where so many of our interactions were relegated to only digital connections, it became all too clear how important the physical and material aspects of our daily lives are. So as we become more and more digital, we actually need to become more and more physically connected with one another, with our products, and with our physical environments. We are now realizing that the most elegant future is when these two worlds are symbiotically connected. Rather than virtual/digital and analog/physical being separate and isolated, we can create this blended and interconnected world of digital and physical materials.

Computing is still inherently physical, just as it always has been. It still relies on precise material properties like conductance, capacitance, density, stiffness, opacity, and even the physical inputs from our own hands. Activating materials, at its core, is a rediscovery of these material properties and an extension of this essential fact. It seeks to imagine and develop different forms of computing that could take all of the awesome qualities of today's invisible digital computing, yet utilize human-scale physical properties that we can see, feel, touch, and play with. We typically think of computers as static and electronic. It's time we adopt a new perspective, moving beyond the boundaries of our computer screens into a vision of computing that places the symbiosis between digital and physical at its center, and brings the capabilities of computers and materials together under one umbrella. How might this change the ways in which we view the world around us?

Material as Computation

It's easy to find examples of physical embodiments of computing if we use the broadest definition of the term: transforming information from one form to another. For

example, a simple mechanism that demonstrates information transfer from a code into a physical mechanism
is the combination lock, which only opens with the right
combination entered into the system. Material computing imagines embedding some of the more complex
properties of today's digital forms of computing into material objects, such as modularity or reconfiguration (like
copy/paste function), or replication and error correction
(as with computer viruses). Conversely, we might even
imagine one day incorporating the unique behaviors of
the physical world (stretchability, bounciness, or transparency, for example) into the cold and static computers
of today. These are just a few imagined realities that may
seem far off but could be closer than we think if we can
take advantage of material-based computing.

One pathway to material computing that has emerged
is called morphological computation, which studies the
relationship between geometry, materials, and computing.[10] This form of computing takes advantage of physical interactions and the interplay of objects to perform
computational tasks. Some of this work emerged in the
robotics communities, where you can imagine trying to
build a walking robot with all of the complexities that
might entail in both the software and hardware. Imagine
designing something that could walk down a ramp on
its own. You might try to design a robotic mechanism
and software program that could give instructions on
how to precisely walk down the ramp. You would need
to program every single motor, gear, and joint at every
single time step. This could quickly get very difficult.
Alternatively, you could design a wheel or some type of
multilegged structure that could step its way down the
ramp with forward momentum, or create some wobbling
structure that moves down the ramp by almost falling,
catching itself, and then falling again. Most of the work
would be done by the ramp and gravity, helping to guide

it from the top of the ramp down to the bottom. This is morphological computation—using physical materials and interactions to solve some of these difficult computational programs. Creating the right physical material structure that *wants* to walk itself down the ramp solves many of the difficult software and electrical problems building the robot entails. There would still likely need to be software to do other things, but the material structures can automatically "compute" some of the mechanics by figuring out how to walk down the ramp, simply through their interaction with one another and the forces in the environment (in this case, gravity).

The notion of morphological computing could be applicable more generally to say that materials and their geometric configurations—how they are put together, with different material properties, or different mechanical and geometric structures—can form the basis of material computing. We can see material computing emerge in various scenarios, some of them for robotic applications, others just about information storage/processing, and others even about physical change, how something goes from one form to another. Material computing allows us to explore all of these scenarios, something that would be challenging with conventional notions of silicon and electronic computing.

As a framework to understand how different forms of material computing compare with traditional computing, the concept of "Turing completeness" is helpful. Named after Alan Turing, a computer is said to be Turing complete or computationally universal if it can simulate any other computer.[11] More specifically, for something to be computationally universal it needs to show that it has both conditional branching (if this, then that) and a way to read and write memory, or to give input, change the program, and get a new output. From a computer science perspective, all real-world computers today are effectively universal, other than the fact that they don't have infinite memory. When we talk about materials as a computational medium it is helpful to think about what type of computational significance they may have. Will these material computers be universal, will they be able to perform all types of computation, like any other computer or programming language? Will they be able to perform other types of computation? Or will they be limited in their capacity? The lock mechanism that we discussed previously has a type of conditional statement—if this combination, then open the door, but if it is *not* the right combination, then do not open the door. The lock does not have the ability to change its own program, however. It can read the program but not alter it; therefore, the lock example would obviously not be a universal computer.

Recently, DNA computing and DNA "hard drives" have been developed that compute much like our traditional computers yet have a number of unique qualities.[12] DNA is considered computationally universal because it can have conditional programming—if these base pairs, then do this, but if these base pairs, do something else—and it has read-write functionality in its program.[13] As we see in mutations and evolutionary biology, the genetic code can be written, changed, and replicated any number of times through the relationship of RNA, DNA, and the ribosome. These mechanisms allow DNA to be read and written and then perform specific tasks or create copies of strands.

Researchers have shown that this proverbial code of life can be repurposed to store and retrieve other types of information, in base pairs rather than 0s and 1s. In his book *Regenesis,* molecular engineer and geneticist George Church explains how he took the entire text of his book, encoded it into a custom sequence of base pairs of DNA, produced synthetic strands of DNA with this code, and then read back the DNA through sequencing to extract the precise text of the book again.[14] Church demonstrated that DNA could be used like a hard drive for information storage and retrieval, and this demonstration has opened up further avenues of inquiry into how we might take advantage of DNA's unique properties for computing.

As one scenario, there's the potential that scientists could create DNA hard drives in the near future that correct errors, repair themselves, and self-replicate. In any one of Church's test tubes there may be millions of strands of the DNA hard drive containing the text of his book. Upon retrieval, he gets not just one copy but millions of copies. And, if a strand has an error in the code, that error can be detected and fixed either *physically*, by changing the order of the base pairs, or *digitally*, by comparing it with the other strands when extracting the information. The text of the book could even self-replicate, producing millions of copies of the same book, similar in many ways to a digital file of the book. This DNA copy of the book could perhaps survive thousands of years, like the DNA we extract from fossils—which is likely much longer than a paper copy or an electronic version. Just think of the fate of CDs, zip drives, VHS tapes, and cassette tapes, all of which remind us of the fragility of our contemporary technologies.

Practically speaking, though, biological computing and information storage may not be seen as a replacement for traditional computing devices; instead, it is the vision to tap into the innate ability of biological materials while enhancing and augmenting their functionality through computation. If we can utilize biological material for computing, then we can allow biomaterial computers to functionally target diseases or cancerous growth. In many ways biological material already does this, but by augmenting their functionality through computational means, we can create networks of logic and sophisticated programs that wouldn't naturally be found. For example, researchers such as Tal Danino have recently been able

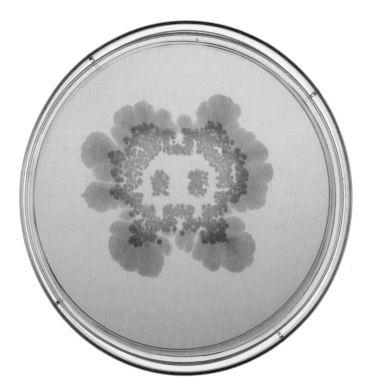

Programmable bacteria printed in precise patterns such that the bacteria can evolve and grow to become collaborators of the artwork. Programmable bacteria are also being used for cancer therapies. *Credit:* Tal Danino

to do exactly this by programming synthetic biological material like bacteria to target and deliver cancer therapies directly within the body.[15] In this way, they are using the functionality of biological material, but augmenting it with new information and agency to create powerful therapies that work with the materials of our body rather than placing foreign devices or drugs in our system. Other researchers have been building biological circuits that embody machine learning while tapping into the biomaterial's ability to learn and adapt.[16] This approach takes naturally intelligent material from biology and combines it with principles from artificial intelligence to evolve more sophisticated programs. These hybrid computational and biological mechanisms could have incredible value in medicine and other biomedical applications. But these approaches also highlight a rapidly emerging example where a different model of computation was created, one that is purely biological-material-based rather than electron- and device-based, where we tap into the material properties to provide solutions that aren't traditionally possible in our classical forms of computers.

As materials are increasingly used for more computational capabilities, they may not always be designed to be computationally universal or the most efficient forms of computing. Some material computers may be slower than traditional computers, or they may have a smaller capacity than today's supercomputers. Yet there are many other reasons why material computing may be a great way to compute. Designers, for example, have been exploring aspects of material computation focused mostly on a material's ability to sense and respond dynamically in its environment.[17] Or as we saw in the case of DNA hard drives, scientists are developing methods where material computation can be more flexible, more adaptable, self-repairable, and self-replicating. For example, if we want to create a computational device to place within the body, it might make sense to trade super high-speed, high-capacity traditional forms of computing for a biological-based computational medium. Some of the traditional devices that we place in the body today—a pacemaker, for example—may start to seem archaic and brutal compared to emerging biological material devices and computers. These biomaterial computers exhibit particular properties that could make them far more robust than traditional devices because the medium allows them to adapt to fluctuations in the body and they don't require traditional electronics, bulky enclosures, or dangerous failure modes. These are properties that are often missing in today's version of electronic devices. These capabilities are not only found in biological materials, however; many innate natural and synthetic materials, such as metals, plastics, and even various fluids, error correct and adapt.

Physicist and bioengineer Manu Prakash at Stanford University has demonstrated that air bubbles can compute digital logic by moving themselves around physical circuits and performing logical operations. Or similarly,

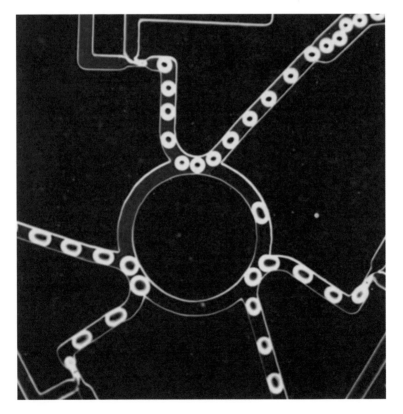

Microfluidic bubble logic, a ring oscillator demonstrating cascadability in fluidic logic. For details, see Prakash and Gershenfeld 2007. *Credit:* Manu Prakash et al.

he showed that droplets of water can actually function as a synchronous and universal computer, rather than electronics and transistors in a circuit.[18] Prakash explains that computing is intrinsically linked to the laws of physics because bits are physical entities and that we can use computers to physically manipulate matter just as we can use computers to manipulate information.[19] They can build all of the fundamental mechanisms of a computer by making any logical operation with hundreds of droplets that move around metallic traces that are roughly the size of a postage stamp. He says, "It's not about manipulating information faster, it is about manipulating matter faster." This work is particularly exciting because it could lead to a droplet-based computing medium where we can build up or change the physical material based on both information and the ambient environment at extremely fast speeds. A change of ambient temperature for example, could cause a transition in both the computing state and the physical output state, altering how information is visualized or how computing translates to human interaction through literally fluid interfaces.

With the example of water computing, and various other forms of material-based computing, the digital realm *is* the physical realm, and our environment influences the way matter computes as well as the physical form that it takes. The power required to compute with materials could be based on temperature or moisture, sunlight, sound, or other widely abundant and underutilized energy sources. In this way, perhaps material computing could one day offer alternatives to our energy challenges and our ever-increasing demand on electric or fossil-fuel energy sources. Or imagine the world's supply of water as a computational medium. Our physical resources could provide ample computing capacity and storage space, and perhaps an abundant free flow of computational power. Maybe these dreams are a bit far-fetched at the moment, but if we can turn any material into a computational medium, it may change the way we see, interact with, and communicate with the world around us.

Communication

I've outlined the ways in which physical materials can compute; communication and connectivity are other key components of any computing platform. Beyond computing information internally, a medium can also pass information externally. One of the ways that digital electronics and computation have dramatically changed our world is through global communication. The fact that we can communicate with another person (or another device) from any point in the world to nearly any other point in a matter of seconds, without any physical proximity or wires, is an incredible achievement. With information and communication happening wirelessly, it certainly doesn't feel physical. Yet information and communication are rooted in the physical.

It's more obvious how materials can communicate locally—two materials in physical proximity can literally push, pull, or otherwise interact with one another, passing information, sensing, and activating each other, to create a communication platform. Think of two people tapping each other on the shoulder and pointing. It is a simple physical movement that communicates information from one person to the other. Or think of pool balls that bounce into one another, passing information in the form of a physical force, causing a physical reaction and translating that information into a new piece of information, the new state of the balls. This simple physical behavior can be translated into information by representing each contact as a 0 or 1, or some type of symbol. An entire game of pool then could represent a poem, or a calculation, or a piece of music. In fact, it has been

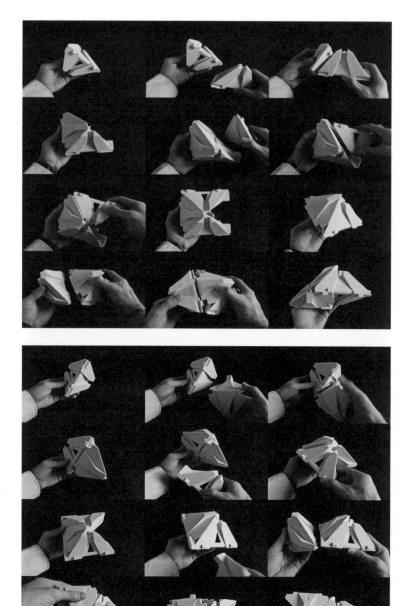

Physical building blocks that embody functional Boolean logic through geometry and the sequence of assembly. NAND Gate top [1,1] = 0 and bottom [0,0] = 1. *Credit:* Skylar Tibbits

demonstrated that pool balls can actually embody digital logic and perform as a universal functional computer.[20]

In my master's thesis at MIT, I developed a series of physical building blocks called Logic Matter, which embodied functional Boolean logic through its geometry.[21] The way that you put the building blocks together would represent 0s or 1s as input; the building blocks would then guide or block future connections and result in a three-dimensional structure that represented a logic circuit or some simple computation being performed. In this way, you could assemble the building blocks based on some predetermined code, or you could use them to compute something almost like a macroscale calculator or abacus, or even store information like a primitive hard drive. Perhaps these examples aren't the most efficient forms of computing, but they demonstrate the translation of information through local, physical, contact.

Material communication is more challenging as we move to a global level, however, since a piece of material sitting on one side of a room may not be able to see, feel, or touch another piece of material sitting on the other side of the room. As a result, we might conclude that it's simply impossible for materials to communicate globally or remotely.

There are two perspectives to consider. The first is that global communication can be a hybrid "digital-to-physical" interface. Rather than thinking of physical material and digital devices as being at odds, imagine them working together. We know that materials sense the physical environment around them (moisture, temperature, light, pressure, and so on) and electronics easily communicate globally, so perhaps they can work together. For example, it's difficult to get a Wi-Fi module on a circuit board to interact physically with the environment or move across the table. And it's challenging to have a piece of raw material communicate wirelessly with someone on the other side of the globe. So it makes more sense to use the Wi-Fi chip for communication, and let the raw material act as the physical interface, sensor, or actuator with the local environment. A simple sheet of wood, for example, could sense the amount of moisture in the environment and activate a Wi-Fi chip to send out communication about the local humidity throughout the day. In this scenario, we can set up a *collaboration* with both the materials and the electronics: humans act as designers, who program functionality, intention, and performance goals; the materials perform in their natural environment, acting as physical sensors and actuators; and the electronic components communicate globally. The collaboration among the three is seamless, taking advantage of their individual strengths.

The second way to consider global communication is to realize that materials already communicate globally if we include line of sight where they can remotely sense, actuate, and influence another material. Recently, it has been shown that plants communicate through physical contact when leaves and branches touch; however, they can also communicate remotely by secreting chemicals in the soil or in the air or by growing fungi off their roots, communicating to their neighbors warnings of things like aphid attacks.[22] Even more surprising, scientists at MIT have recently developed a way for plants and humans to communicate whereby spinach plants can sense and detect explosive devices and then emit a fluorescent signal that can be detected with a device for humans.[23] In this way, the plants can sense and communicate not only to themselves but also to the outside world. Ants, slime molds, and many other species communicate through distributed chemical signals rather than through auditory signals like most human communication. Humans and many other species also communicate through nonverbal/auditory pathways, as seen in sign language, body language, facial expressions, and physical contact. There are many ways to create both local and global communication. In fact, before today's modern forms of communication through landlines or wireless signals, humans communicated through many remote techniques such as smoke signals and Morse code, which transferred information across long distances and translated it into physical patterns, visual graphics, and ultimately communication. We can similarly utilize these simple, yet sophisticated physical techniques for local and global communication in the design of material communication.

Recently, there have been a few examples where researchers have developed remote communication through simple material behaviors. A team of scientists created a printed lattice structure that functioned as a morphable antenna. They could change the resonant frequency of the antenna by changing the temperature of the environment, therefore controlling the sending and receiving of information between two points across different frequencies.[24] As an everyday example, think of a speaker and a microphone: they are the same mechanism but with reversed functionality—a speaker turns electricity into sound, and a microphone turns sound into electrical signals. Both the speaker's sound and the microphone's signals happen as a result of a physical material—a thin membrane that vibrates to produce or receive sound waves. The vibrating physical material can produce sound, transmitting a signal from one location to another, which is received through the vibration of another material. To continue the sonic analogy, a vinyl record conveys information through physical geometry, bumps and grooves, which store the information of a song and translate these geometric complexities into beautiful sound through physical, vibrating speakers.

As a thought experiment, if we had a thin film that could be activated by infrared (IR) light, causing it to undulate, we could use this to create semiglobal communication. We could place a piece of this material on one side of a room and subject it to pulses of IR light, on/off patterns, causing it to move. This would behave like the speaker, creating a vibration. On the other side of the room we could place another piece of the same material, acting as the microphone, which would receive the vibration of the speaker. This would cause the microphone material to vibrate, which could then be read out as information. The frequency of the vibration in the microphone material would relate to the activation from light on the speaker material. This hypothetical example demonstrates how two physical materials could communicate and create a wireless IR gauge showing that materials may be able to communicate in unusual ways and translate information from one location to another. This example is a reminder that our current means of information and communications are actually quite physical, and with a more careful lens turned to investigating the inherent capabilities of materials, we can create physical-digital hybrids. These hybrids serve as models of computing and communicating platforms that create elegant solutions by tapping into material properties.

Computing without Efficiency

Some computer scientists argue that the fundamental aspect of computing comes down to what can be (efficiently) automated.[25] But what if computing were not only about efficiency, automation, or optimization? What if computing were about inefficiencies, creativity, and all of the other messiness in our lives? Why did computing get relegated to being only pragmatic, optimal, or efficient?

The first applications of desktop computers were thought to be spreadsheets and accounting, but luckily that wasn't the end. Personal computers quickly came

along that began to address all of our weird and personalized interests through today's music, photography,
gaming, video editing, apps, and so on. Perhaps the most
interesting emergent domain in computing today is the
creativity and aesthetics of artificial intelligence (AI) and
machine learning. For example, Google's image generator created unimaginable scenes, ones that humans
could hardly dream of creating, rather than optimizing
or solving problems. We should be promoting computation for creativity and its generative qualities not only for
optimization. Efficiency and automation invoke speed,
performance, and optimal solutions. At the other end
of the spectrum, however, there is creative computing,
in which we can explore the intersection of the creative
arts, design, and computing. Seymour Papert, Mitch
Resnick, Muriel Cooper, John Maeda, and many others
have pioneered playful computing and creative coding
in an effort to promote curiosity and exploration.[26] In
Processing, a software language developed by Ben Fry
and Casey Reas, each file is called a *sketch,* and it is precisely the idea of creatively *sketching* with code that is
desired—making mistakes, being surprised, discovering
something, rather than optimizing or trying to find a singular, efficient solution, as the primary goal.[27] Computation itself can be used as an idea generator, just as a quick
sketch allows for misinterpretation and creativity, or as
watercolor makes it possible to blend and blur the lines
between detail and fuzziness. Creative coding allows for
new ideas to emerge.

Similarly, by computing with physical materials, we
can embody these *inefficiencies* of computing and put
them to creative use. Materials have unusual traits that
may not increase speed or capacity. They may not be
faster than silicon at flipping a bit and may not have
unlimited storage capacity (although DNA hard drives
might get pretty close in some aspects). Materials are

traditionally nuanced, weird, and dynamic; we should tap into these traits and do something interesting with them.

With programmable materials, we can make materials active: we can build a simple set of building blocks that can be combined to make materials behave in surprising ways—sometimes with higher performance, sometimes with storage or efficiency in mind, and at other times with creativity, playfulness, or lifelike qualities in mind. Materials can also embody polymorphism to its fullest. Polymorphism, a term used in both biology and computer science, is about a single code that can manifest as many different outputs.[28] This creates a very interesting conundrum: if you have a linear set of instructions (like an algorithm written in code), or a set of physical behaviors in a material (sense, actuate, fold, curl, twist), or a series of choreographed rules for a dance, or a procedure in Brian Eno's algorithmic music, or the instructions for realizing a Sol LeWitt artwork—how is it possible that unexpected things can emerge?[29] If the rules are well understood, and the operations or procedures among those rules are clear, then how could anything surprising happen? Polymorphism is one way to explain that. Polymorphism is the ability for difference to arise from the same embedded information. Polymorphism insists that different results can emerge, even with the same information and the same procedures. And this is the surprising and awesome place that materials can take us in computing.

As a thought experiment, imagine that we can create a material composite that has certain parts activated by moisture and other parts activated by heat, in an alternating pattern, heat-moisture-heat-moisture-heat-moisture. When the moisture zones are activated, they will bend so that the overall structure makes a square with four corners. If the material is heated (without moisture), however, the previously straight segments will curl in the opposite direction, creating a circular shape. And if we activate this material structure with *hot water*, it will do something completely different—the moisture zones will bend downward, the heat zones will curl upward, and you'll get a sinusoidal-type shape. For example, if we put the wet material on the bottom and the heated material on the top, it will behave one way with the sun shining from above, and in a completely different way if you set it in a puddle. Both the physical structure of the material (where the materials are in relationship to one another) and the changes in the environment (different amounts of energy, different times of activation, and so on) will conspire to create different results even though we are using the same material composition. The executed *program* comprises the material's geometry and the patterned placement of the material properties. This leads to a logical operation for the material: If water, then do X, or if heat, then do Y, and if water AND heat, then do Z. If we make even more complex material structures based on this with fluctuating physical environments, we could get extremely sophisticated, complex, and surprisingly brilliant material computers that aren't just about efficiently solving a problem or automating a human task. They are about discovery, design, or performance.

Another way to think about creating differentiation through simple rules is to consider the physical embodiment of chemical morphogenesis, a topic that fascinated Alan Turing and was the focus of his work just before he died in 1954.[30] Morphogenesis is the differentiation of chemical, biological, or physical processes, which can grow from homogenous building blocks (such as cells) to complex patterns like zebra stripes, cheetah spots, or all of the complexities of us humans. These patterns emerge from instabilities triggered by random disturbances in the homogeneous system, activating and deactivating the system. The instabilities and random fluctuations of our physical environment combined with simple material-based rules can lead to incredible complexity. Entire fields of study such as chaos theory and complex systems focus on exactly these types of complexities that emerge from simple rule sets. Complex systems are often defined by very important initial conditions (such as the genetic code or programmed behavior) and a feedback mechanism, both of which can lead to dramatically different outcomes from even simple changes in the environment. The common thought experiment is a marble that is dropped precisely on the tip of a mountain. Very minor changes on one side or the other side of the mountain ridge will lead to dramatically different paths for where the marble will end up at the bottom of the mountain. Even with the same marble and the same mountain, minor changes can create very different outcomes.

Our physical environment is a perfect petri dish for a computational material medium to flourish, offering complex and creative outcomes. This possibility is ripe for use in design and engineering. Designers can imagine a product made of coded-material rule sets and with random fluctuations in the environment: a new pattern or structure could emerge—like a chameleon—ever-changing yet always made of the same physical material.

In cellular automata—a classic computational model that has white and black colored cells with embedded rules for their color based on their relationships with their neighbors—researchers have studied how patterns emerge from simple sets of rules. In his book *A New Kind of Science*, Stephen Wolfram outlines four types of patterns: growth/death, where patterns either take over or completely die off; repetitive patterns endlessly making the same thing; chaotic patterns that have no discernable structure and look like pure noise; and, most interesting, patterns that oscillate between chaos and order.[31] I believe it is this fourth type of pattern that we should strive for when working with programmable materials. We want to build active systems by embedding simple rules into materials, enabling them to transform themselves, interact with one another, and enhance their function, performance, or aesthetics. We want them to enable truly unique and surprising results even from simple embedded rules. These capabilities shouldn't be just repetitive tasks, or behaviors that eventually die off; nor do we want them to be random glitches or spasms of behavior. We want interesting and useful behaviors, much more like the patterns of human or natural systems that are both chaotic and repetitive in all of their delightful and surprising ways.

Order from Chaos

MOST MANMADE THINGS seem to fall apart. Beautiful new products eventually deteriorate, buildings need maintenance, food rots, clothing rips, cars eventually die. It is sad how things typically move from order to disorder. Yet, nearly everything about the natural world starts out by growing from seemingly nothing into something amazingly complex, functional, self-repairing, and self-replicating. All of life strangely appears to resist this, building order from disorder. As the second law of thermodynamics states, the entropy of an isolated system must always increase (or sometimes remain constant). Yet many things seem to go against entropy and scientists sometimes even describe life as the fundamental pursuit of negative entropy. Most systems that appear to lose entropy and create order from disorder are not isolated systems; there are fluctuations in the environment or the amount of external energy input, or we can even design the components to promote their self-organization. Globally, however, the environment surrounding these systems must increase its entropy at least the same amount, in which case the global entropy never decreases. In everyday terms, what is amazing is that we can see spontaneous patterns emerge in crystallization, or swarms of insects, or a single cell growing into a human that can regenerate and self-heal when injured. The question is why this happens and how to utilize it.

Active matter can allow us to tap into these incredible behaviors often only believed to be possible in biological or chemical domains. We can imagine a near-term future when everyday physical materials will build themselves, grow, adapt, transform, and improve over time. The property of self-repair, for example, no longer applies solely to living things; it is starting to be embedded within our built environment. Researchers can now create self-repairing materials like self-healing concrete, polymers, and composites.[1] As we will explore throughout this chapter, many examples are emerging with animate material behaviors in traditionally static material systems.

There are more precise descriptions of entropy that can help explain this complex phenomenon and the process of growth and decay. The most common illustration of entropy involves two chambers with different temperatures, one hot and one cold. When the two chambers are allowed to mix, the entire system will move toward an equilibrium temperature. In 1867, James Clerk Maxwell described a thought experiment as a hypothetical violation of the second law of thermodynamics that describes entropy, now called Maxwell's demon, where a small door could be opened and closed quickly to allow only the fast molecules (in this case, hot air) to enter into a second chamber. This process would separate the two chambers into one hot and one cold, effectively reversing entropy by creating order and seemingly violating the second law.

In Maxwell's chamber, after the two sides are allowed to mix with hot and cold molecules bouncing around the tank, it is theoretically *possible* that for one moment in time, all of the hot molecules would be momentarily on one side of the chamber and all of the cold molecules on the other. It is possible, yet incredibly unlikely—because the most likely scenario is that the two types of molecules will move to be a mixed distribution of hot and cold molecules at any given time. So in order to make the opposite happen, we need to coax the molecules to make them comfortable in the seemingly impossible condition where they are separated. One simple example where we can manipulate the local environment (seemingly violating the second law) would be to use the vertical axis to separate temperature zones. We know that hot air rises, so we know that the hot air molecules will rise to the top of a space while the cold air molecules will tend to stay toward the bottom. If we were able to design the chamber in just the right way (making it very tall, for example), or if we were able to rotate the chamber in different orientations, or if we could add a heat source outside the chamber (all of which emphasize that this is not an isolated system) we could promote the hot and cold molecules to separate above and below one another.

Another way to describe entropy is a system seeking equilibrium: any object or system, will tend to move toward a position that is the lowest-energy state. For example, a ball will roll down a hill, moving from a higher position with greater potential energy to a lower one with less, and then naturally come to rest in the lowest-energy state. Yet, as in the example of a Rube Goldberg machine, you can design the "hill" (or machine) in such a way that the ball can do extraordinary things like fly through the air, or perfectly balance on top of objects, fall into precise locations, or set off entire chain reactions, seemingly on its own. Or if you imagine dropping a number of pieces of paper, they will likely scatter into a random pile on the floor. But if you could fold each piece into a uniquely designed paper airplane, you might be able to drop them and have them fly into very specific and surprising patterns. I like to think of this as designing the system so that the lowest-energy state is a very useful and interesting one. In this way, the equilibrium state does not necessarily mean a destructive, degenerative, or disordered state. Instead, think about how we can use the concept of entropy to design non-isolated systems that move themselves toward states that get better with time. The main challenge is how to create the lowest-energy state to become the most functional, ordered, or even surprising condition.

The phenomenon of self-assembly can be described as individual components that spontaneously assemble ordered structures without human or machine intervention. This is the underlying principle within biology and chemistry that accounts for everything from how humans are built from DNA or how ice crystals are formed from water to how planets are formed in the solar system. We can think of self-assembling systems as individual parts moving toward a final configuration, or an equilibrium point. This is slightly different from self-organizing systems, in which components don't necessarily move toward equilibrium, but can move between multiple states, oscillate, and may never come to rest in a final configuration. Schools of fish, flocks of birds, sand ripple patterns, and traffic jams that grow, organize, change, and dissolve repeatedly are all examples of self-organization.

In biology, self-assembly is the primary approach for construction. There aren't many other techniques to build with biological material, given how small and complex the environment is. DNA assembles itself through complementary base pairs, proteins fold themselves, and higher-level molecular structures come together to make functioning capsids like viruses or other biomolecular structures where order and functionality build themselves. Recently, scientists have been able to tap into this phenomenon and design DNA structures that can self-fold and self-assemble into nearly any two-dimensional or three-dimensional shape. Peng Yin's group at Harvard's Wyss Institute has pioneered an approach now called DNA origami to use DNA as a building block, much like a Lego, for construction at the nanoscale.[2] With this technique, many other researchers around the world can now program DNA with specific base-pair sequences to promote self-assembly, forming precise nano-architectures from 2D and 3D bricklike modules. They have demonstrated hundreds of 2D shapes such as letters of the alphabet, emojis, and symbols. Three-dimensionally, they have also created hundreds of demonstration objects using a canvas of DNA building blocks to create volumes like DNA spaceships, drug delivery capsules, and other geometry.

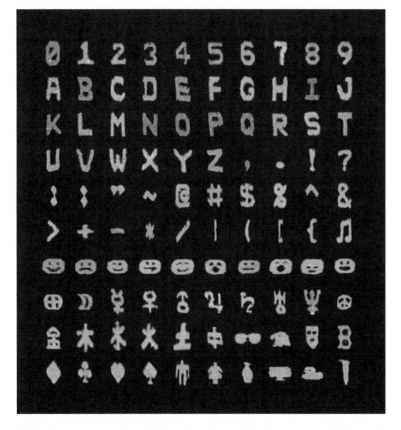

DNA origami structures formed from the self-assembly of programmed DNA strands forming precise nano-architectures. *Credit:* Wyss Institute

This approach may very well translate into bottom-up assembly and manufacturing approaches for various applications of nanotechnology in the near future.

To create self-assembling systems, we first need to understand why self-assembly works and how things move toward equilibrium. Then, we need to find ways to take advantage of this equilibrium seeking to promote order out of chaos. At the Self-Assembly Lab, we define self-assembly through its three core ingredients—energy, geometry, and interactions.[3]

The first ingredient we look for in any self-assembling system is energy. The energy imparted into a system needs to be just right to promote assembly, and the amount of time or frequency of this energy activation needs to be designed to promote the structure to easily find equilibrium. It is also important to think about where the energy is coming from—where you can find abundant energy sources, whether that is vibration in a system, or changes in temperature, pressure, or wave energy—these are often common and overlooked sources of energy that we can work with when designing a self-assembling system.

A number of years ago, the Self-Assembly Lab developed a project in collaboration with the molecular biologist Arthur Olson, around a macroscale and tangible demonstration of self-assembly.[4] We based the project on a poliovirus, a tobacco plant virus, and other biomolecular structures that naturally self-assemble. We produced a number of glass containers that had simple plastic parts inside. When you shake the container, the parts inside spontaneously come together, building order from disorder. If you do not shake the container hard enough, the parts don't have enough energy to come together. Conversely, if it's shaken too hard, the parts will separate themselves. If you try to shake it with intention, almost like a puzzle or game bouncing them into place, it is often worse than if you shake it randomly. The trick is

A tangible demonstration of self-assembly based on bimolecular structures like the polio-virus or a tobacco plant virus. When the container is shaken with the right amount of energy, the parts inside spontaneously come together, building order from disorder. *Credit:* Skylar Tibbits, Arthur Olson, and Autodesk

to find just the right amount of energy so that the parts can move around freely, find one another, and connect. This simple experiment demonstrates that physical objects can easily transition from order to disorder and back again when given just the right external conditions.

The perfect amount of energy for self-assembly is similar to Brownian motion, where the components are able to move around, bump into one another, and connect easily. This amount of energy input can vary from system to system, depending on the given environment, material property, and bonding characteristics. For example, the amount of energy and the type of energy needed to move objects underwater is very different from objects tumbling around in a container or flying around in the wind. Underwater you may need waves, air pumps, or propellers to produce turbulence in the water, whereas with a tumbling system you may need a motor to drive a rotating chamber. If the components underwater are neutrally buoyant, they will need a smaller force to move them around the tank, whereas if they are made of metal and sink to the bottom, they will need a much greater force to keep them moving freely in space. In a tumbling scenario, if the components are made of rubber versus clay, the energy landscape will look very different. The rubber materials will have a very delicate balance of forces because if the forces are too strong, they will bounce off one another, while the clay components can easily stick to one another or mold around one another. Finally, the bonding strength and property of the connections can change the amount of energy in order to promote successful connections and weed out incorrect bonds. If the connections have significant bonding strength, then you will need less energy to promote the connection of the units, but you will need greater energy to ensure that incorrect connections break off and even greater energy to disassemble the entire structure.

The amount of energy input is subject to a very delicate balance: if there is slightly too much energy, the components can bump into one another with too much force and either bounce off one another or break apart existing connections. With too little energy, the components won't be able to find one another and will not move around and connect with one another. Similarly, there needs to be enough energy to break apart incorrect connections, yet not so much energy as to break correct bonds. It is like the Goldilocks principle for activation energy in any self-assembling system—with just the right environment, the structure moves toward equilibrium.

The second ingredient is the geometry of the physical parts in the system—its individual components, materials, and connections. For living systems, the components may include DNA, proteins, cells, and other materials, all of which have physical properties such as size, shape, density, and bonding strength, which will influence their

A series of 36" diameter weather balloons filled with helium inside a fiberglass frame with Velcro nodes. The balloon structures float around in the courtyard and self-assemble into various lattice structures. *Credit:* Self-Assembly Lab, MIT, Autodesk

The balloon structures self-assembled into a large-scale cubic lattice. After the helium fades, the balloons float back to the ground and the self-assembled lightweight structural lattices remain. *Credit:* Self-Assembly Lab, MIT, Autodesk

interactions with one another. The geometry of a component is obviously important because it has to effectively interact with surrounding components and easily come together to build a precise structure. Certain geometries will promote two-dimensional structures, and others will promote three-dimensional structures. Some geometries will promote the linear growth or aggregation of materials, like lattices, and yet others will promote the formation of closed-loop structures. Think of lipids and bilayers that are formed from hydrophobic and hydrophilic interactions. The geometry and orientation of the lipids create very different structures in different environments. The key is to craft a local geometry that will aggregate to achieve the desired global structure.

The size and density of the materials influences how they mix or separate when we add energy. If they are all similar, the components will mix in a uniform manner, but if they are vastly different from one another, they will separate into different regions. This can be seen in a principle called granular convection, and it comes into play when we're separating elements into constituent parts.[5] A coin-sorting machine is one example—coins separate due to their differences in size and density. Conversely, we can use this principle to promote close proximity and increase the chances that components will assemble on their own. For example, mixing materials with different densities in a tank of water will cause some to rise while others will fall. This rising and falling behavior could enable objects of similar density to connect with one another. Some parts that are very buoyant might assemble into 2D sheets on the surface of the water and less buoyant parts will assemble at the bottom. Designing their interactions requires close attention to a material's characteristics such as elasticity, stickiness, or friction, which will dictate the types of interactions that occur among the components themselves.

Finally, after ensuring that a system has just the right amount of energy and the components are designed to promote assembly, you have to take a look at the connectivity and interactions of the components. The goal of any self-assembly system is to promote the precise assembly of something useful, such as a product assembled from a number of pieces—not to just end up with a random mess of parts. The stickiness between parts is important, and this can be created by using adhesives, Velcro, magnets, surface tension, Van der Waals forces, or various other approaches to creating physical connections. If *everything* is extremely "sticky," however, the system wouldn't deliver an ordered structure. The components need to have just the right amount of stickiness, in just the right places. The strength of the connection is important as well. If it's too strong, the components will stick but won't be capable of breaking apart. This sounds like a good idea, but what happens if the parts come together in the wrong way? They will be forever stuck in the wrong place. If the strength of the connection is too weak, however, the parts will always fall apart, even when connected correctly.

With just the right amount of connection strength, the correct parts can connect, stay connected, and continue to grow stronger, while the incorrect and weaker parts will fall off. In other words, the parts themselves correct their own misguided attempts at connecting. For example, if there are many "male" and "female" parts tumbling around in a container, the similar connections (male to male or female to female) will have very little or no strength when they meet one another, whereas the male to female connections will be much stronger and will remain connected as the part continues to tumble. This is a simple technique to design error correction into a system based only on the strength of the connections.

Patterned connections are another example of error correction. One way of thinking about this is a

A number of unique components tumble around within a tank of turbulent water, eventually self-assembling into a chair. This process was designed by using custom geometries and nodes that encode the correct assembly sequence. The process was filmed over seven hours, after which the fully assembled chair was complete. *Credit:* Self-Assembly Lab, MIT

lock-and-key type of mechanism. A lock-and-key type joint is useful because it can dictate that only the exact components can connect with one another, while all other components can't—they simply don't fit geometrically. This is particularly useful when you're designing structures that have a number of unique connections. You can create an infinite number of geometries that only fit with the complementary pair. Just like car keys or house keys that have many variations, you can create a number of unique connections that allow for only complementary connections. The lock-and-key patterning offers a wide variety of possible combinations (limited only by our geometric imagination) and can allow for geometric error correction. We demonstrated this feature in our Fluid Assembly Chair project, where unique components had lock-and-key connections that promoted the self-assembly of a three-dimensional chair when tumbling around in a water-filled tank.[6] Similarly, we showed the successful self-assembly of a cell phone from a few simple building blocks, the front enclosure, the rear enclosure, and the circuit board/battery core. These components were designed with precise male/female connectors and a series of polarity patterns that promoted the complete self-assembly of the functional phone after being tumbled in a mixing chamber.

Another example of patterned error correction comes in the form of polarity, which can be magnetic polarity, polarity in static charge, hydrophobic or hydrophilic polarity, or a variety of other types. Magnetic polarity is the most common: a positive side attracts a negative side and vice versa, whereas two similar sides repel each other. This allows for the precise patterning of components, connecting only complementary polar neighbors. For example, a string of magnets in a positive-negative-positive-negative pattern will attract a complementary strand (similar to DNA, yet DNA has four base pairs), with the pattern negative-positive-negative-positive. A similar patterning technique can be used with surface tension, for example: a hydrophobic material will repel a hydrophilic component but attract similar materials to itself. The "Cheerios Effect" is one example of this concept: Cheerios tend to cluster into hexagonal patterns in your cereal bowl due to the surface tension created by the milk and the circular shape of the cheerios.[7] To take this one step further and start designing things that come together on their own, it may not be enough to simply use polarity because there are usually only two poles (positive and negative), and in some cases a third, neutral (like a piece of steel that can connect to either magnetic pole). Wish as we might for an infinite number of poles, they don't normally exist—unless we design them!

Self-similar modules were released in a 500-gallon tank of turbulent water and self-organized into lattice structures based on the interactions of the elements and movement of water in the tank. The complete structures can then be removed or disassembled and thrown back into the chamber to self-assemble once again. *Credit:* Self-Assembly Lab, MIT

By designing patterns of complementary matches, we can actually create an unlimited number of "poles." Take small magnets, for example, that only have two poles. We can place them in a pattern with positive and negative orientations on a surface. Then, we can design a pattern of positive/negative pairings on another surface that matches the first set perfectly. If we change either the polarities or the location of the magnets, we could promote or inhibit the connection of units indefinitely. With this technique of patterning, we can create an infinite number of poles simply by designing patterned stickiness into our components. With recent fabrication techniques, we can even print custom patterned magnetic sheets that have a limitless arrangement of poles and complementary patterns.[8] In DNA, the complex pattern of ACTG helps to promote the successful pairings of DNA strands while weeding out errors in our genetic code.[9] We can do the same with a series of polarity patterns on a component.

We can see how it is feasible to promote the self-assembly of structures by building order out of chaos. We can specifically design around energy, geometry, and smart interactions to ensure these products move to better states and better functionalities. Collectively, we have gotten used to the idea that all manmade things will eventually come to an end. But perhaps we can shift our perspective. Theoretically, we should also be able to design a product that resists destruction, natural degradation, and obsolescence through error correction. But rather than fighting the degradation, why not design and manufacture structures that are reconfigured, converted, or altered into other products when they aren't needed anymore? Every last component, material, and connection could be used for something else, each time bringing along a little bit of knowledge from its previous form, continually seeking improvement.

Less Is Smart

MOST PRODUCTS, just like pieces of infrastructure and much of the manmade world around us, have one thing in common: they are designed to be stable and static; they are engineered to fight *against* all the forces around them—gravity, vibration, temperature, moisture, and so on. They are designed to be *robust*. They are generally *not* designed to be lean and adaptive, or flexible, or reconfigurable. Today's products often don't take advantage of their material properties and aren't programmed to have any of the lifelike qualities that are possible with active matter.

We compensate for the lack of adaptability or lifelike qualities in our products by creating so-called smart versions of them: smart thermostats, smart clothing, smart shoes, smart cars, and even smart bassinets that sense babies' sleep patterns and adapt the sounds or motion accordingly. These "smart" products often get more expensive, heavier, and more complicated to build, however. They become easier to break and more difficult to use, they consume more power, and they are overwhelmed by more mechanical and electronic devices. So how do we get the smarter products without the extra components, cost, and complexity? And how do we get better products for our ever-changing needs without resorting to standardized solutions like the generic version or the super complicated approach where we throw all the bells and whistles at it?

Our goal should be to make active products, by which I mean products, objects, or materials that are literally active—they can move, reconfigure, transform, assemble themselves, or adapt to their surroundings. In order to achieve active products, we need to reconsider the way we think and talk about our (statically designed) world. One of the fundamental principles of engineering has always been that any product or system ought to be designed to resist the forces that may lead to its destruction—in other words, *designing for robustness* in the traditional sense. The result is that systems tend to be overengineered, and intentionally so. For example, various safety factors exist in buildings, bridges, cars, or planes to ensure that structures will withstand more than the weight we anticipate they will bear. Of course, this is extremely important for safety. But from a materials perspective, it's wasteful. Perhaps it's time that we rethink or expand what we mean by "robust"—and redefine "smart" in the process, too. A structure that is robust could also be active, lean, adaptable, and error correcting. A number of researchers have built morphing, self-adapting bridges and slab structures that can change dynamically as load is applied.[1] These structures, although currently built electromechanically, demonstrate extremely lightweight and more materially efficient structures that can span and cantilever significant distances. They are one step closer to this dream of higher-performing structures with minimal materials while adapting to complex dynamic situations—without more components, more material, or more rigidity. In the end, *less is smart:* the more we can do with less, the smarter our systems will become.

Error Correction in Products

The principle of error correction, which we touched on in the last chapter, is critical to creating active products and structures. It allows us to ensure that accurate products are assembled in the factory and it can also inspire us to try to design structures that improve over time. Like a fine wine, well-worn baseball glove, a cast-iron pan, or a pair of jeans, some products *do* get better with time. Similarly, much like DNA, physical materials can actively error correct by continuously checking for changes in the environment or their internal structure, and then self-heal, adjust, or tweak themselves if needed. We can actively engage, enhance, and make the most of this principle. Now the challenge is to figure out how to actually design for this to happen in the products around us.

We can start by looking for timeless design and material functionalities. Design sometimes enhances this timelessness—think of classic furniture, vintage cameras, or classic cars. The designs of these products have lasted through time and often still look as radical yet elegant today as they once did. Materials and functionality can last as well—and even get better. Concrete, counterintuitively, is a material that can get stronger with age due to the hydration process and interaction of the material elements.[2] We can imagine designing systems of all sorts—manufacturing, products, or environments—where we impart energy and just the right conditions to promote error correction and overall improvement over time.

There are ways to improve a design to make objects more robust and less static. Natural systems, like our bodies, plants, animals, chemical systems, and many others, exhibit characteristics of robustness and they are resilient—they are lean, soft, and agile, and can adapt to changes in their environment. These systems resist failure very differently compared to the ways we typically engineer systems. For example, bone grows with variable density and stiffness depending on its location in the body and the weight an individual carries, both literally and figuratively. An astronaut's bones will adapt and reduce their mass, and then regrow when they get back to earth. Many natural systems, including our bodies, can regrow, adapt, and correct errors when needed. In other words, error correction itself is a form of robustness.

To understand how error correction can work in everyday objects, let's look at a simple example of building a circle. One manufacturing approach to building a circle out of components is to make the parts with extreme precision, with the exact angle needed for the exact number of parts. If you start to connect the parts with rigid and strong connections, you will need to build in some tolerance. If the weather changes, or the moisture or temperature increases, the parts may become slightly larger or smaller than originally designed. The machine that was used to fabricate them and the materials themselves all have some amount of tolerance. If you have just a few parts, it might work. But as you increase the number of parts or tolerance propagates, they won't come together to form the perfect circle. It is likely that the last part won't fit perfectly to complete the circle (it will either be too large or too small). Even if the parts fit or can be forced together, fluctuations in the environment may create differential expansion or contraction, causing the circle to buckle or bulge. Every piece has only one place in the finished circle, and it needs to be made with 100 percent precision to create the perfect circle.

A second approach to building a circle demonstrates error correction in the form of flexibility. If you build each part so it can pivot or flex where it meets the neighboring part, you can build error correction into the system. With flexible connections, as the parts of the circle come together, they adjust to one another. As the last part goes into the circle, all of the others can adjust their angles to create perfection. Adding simple flexibility in the connections allows the circle to find its own equilibrium. When the environment fluctuates, these units will adapt and adjust, always maintaining the perfect circle. Flexibility serves as a form of error correction that provides us with more robust structures without adding more material or complexity to the design.

Similarly, when we are assembling something with bolts, we are often told not to overtighten the first bolt. Rather, we hand-tighten all of the bolts and then go back around and tighten the rest. This ensures that all of the bolts are tightened evenly and are well aligned. Or when bolting a tire onto a car, it is recommended to follow a star pattern of tightening to ensure that the tire sits perfectly snug. If you overtighten one side substantially more than the other, then the initial side will be tightened off-axis. These simple techniques allow structures to have some flexibility and self-align to fall perfectly in place without measurement or precise machines.

Redundancy

I've argued that we should aim for less material and less complexity in our designs, but there is also a case for material redundancy if we view the term in a different light. If you have more material than necessary, you can sometimes create a very simple system that may be fast, inexpensive, and easy to build. Think of a bird's nest—it can be expediently assembled, often with a lack of precision. The geometric intricacy of the nest can be robust and create breathability and flexibility—features that a more rigid structure may not achieve. So we can sometimes counteract a lack of efficiency in material usage if we can increase speed, improve material placement, and decrease cost with simple and imprecise components to create a robust and adaptable structure.

For example, one project that highlights this principle of material redundancy and adaptability is one that our lab developed in collaboration with Gramazio Kohler

A four-meter-tall tower built with only loose rocks and string using the principle of granular jamming to create load-bearing structures that are fully reversible. *Credit:* Gramazio Kohler Research, ETH Zurich, and Self-Assembly Lab, MIT

Research at the ETH Zurich, where we created a system of "granular jamming" that uses rocks and string to create load-bearing columns or walls.[3] Granular jamming is a material phenomenon that allows disordered particles to transition from a liquidlike state into a solidlike state and back again. Think of coffee in a vacuum-sealed package: that package is typically very stiff and feels like a rock. But when you open the package, it easily flows out when you turn it upside down. We took advantage of this principle; however, we developed a granular jamming system that doesn't require a vacuum or a membrane. Given that the membrane is susceptible to puncturing and the vacuum is energy intensive, we wanted to find a new technique for granular jamming in order to use it as a construction method. In order to create the jammed structure, we create an elegant balance of forces by depositing the right mix of loose rock and continuous string, layer by layer, within a bounding box. After a layer of rocks is poured into the box, a robot is used to unspool a series of loops of string, then another layer of rocks, then string, and so on. When we remove the bounding box, only the rocks that were near the string get stuck, while the rest of the rocks fall away. The rocks can't go anywhere when the bounding box is removed because the rocks take the compressive forces, and the string takes the tensile forces, which makes the structure jam into a solid object. This technique creates a load-bearing structure without using structural members, connectors, adhesive, or other binders.

Our most recent approach to this research continued to advance our goal of tapping into the phenomenon of granular jamming and letting the material do the work, making it as easy and fast as possible to build.[4] In this latest version, we used a simple unspooling technique whereby a spool of string will uncoil itself into perfect circles, depending on the spool and the height above the ground. This method uses the string approach, with

A zigzag wall was built with only loose rocks and coconut husks using a slip-forming method to promote granular jamming in a fast, efficient, and reversible manner. *Credit: Self-Assembly Lab, MIT, Google*

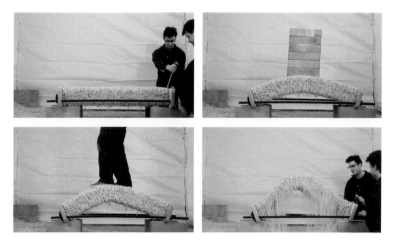

A horizontal beam was built with loose rocks and string using the principle of granular jamming. The beam was then compressed into an arch and repetitively loaded until collapse. This showed the morphability, load-bearing capacity, and reversibility of granular jamming. *Credit: Self-Assembly Lab, MIT, Mechanics of Slender Structures Lab, Boston University*

off-the-shelf spools of string, but eliminates the precise robot deposition, replacing it with a simple principle of physics. Let the string make precise patterns on its own. We then simply poured the rocks and unspooled the string to make columns and walls.

Through granular jamming, structures can actually become stronger with load because the rock and string increasingly act more like a solid. We realized that if we used a top and bottom plate and compressed them with a threaded rod, we could jam them into solid structural components and move them around. We built a column and then rotated it into a beam or a bridge, as well as a wall, and then rotated it into a slab, and then walked across the beam and slab structure. They behaved like a solid structural member, but once we removed the top and bottom plates, the structure would instantly dissolve. This meant that we could switch the structure on or off at any time. We could build extremely fast, make them load bearing, and then instantly switch them off, falling away into a pile of rocks and string.

Going one step further, we realized that if we continuously compressed the horizontal beam, it would start to morph, like a semisolid material, into an arch, which we walked across and loaded at various points. This exploration showed the fascinating and strange ways that simple materials like rocks and string can behave: they can act like solids, semisolids, liquids, and even switchable devices with reversible properties. We can make extremely strong structures with minimal construction time, or soft structures that can be sculpted into shape.

Each of these granular jamming techniques work only because of the redundancy of the material system. Since it's not feasible to place and position every single rock (there were hundreds of thousands of rocks in the experiment), we can't be certain that the connection of the rocks or the location of the fiber is perfect. We can, however, employ rock and fiber in just the right amounts to ensure the stability of the structure. This type of redundancy can also make a robust system, even though we have very little control over the precision or the details. So rather than robustness being about more control and attempting to fight the forces of failure, as is the typical case with a structural beam or component, robustness can also be achieved through expedience of construction, more material yet less control over their placement. In this way, the system is working in harmony with the forces of compression and tension to get stronger. We are certainly using more material than necessary, but the construction process is far faster than if it involved manual placement or poured concrete. In other words, we can sometimes gain speed or performance by letting go of control in the process.

Active Products

With material capabilities and fabrication processes advancing rapidly, research teams are increasingly demonstrating a new class of products that are no longer static and passive. At the Self-Assembly Lab, we have created a number of examples of active products like a flat wooden sheet that jumps *into* a table, assembling itself from its flat-packed box; or a shoe that forms itself, eliminating molding or manual forming in the factory; and a knit garment that adapts to the shape of the body and changes porosity and thickness to keep you comfortable in any environment.[5]

Underlying many of these examples is a technique we developed for transforming flat sheets into three-dimensional shapes. To do this, a piece of stretchable textile, like Lycra, is initially pulled tight and wrapped around a plate. The prestretching process embeds and stores energy in the material, to be released later on. The pattern of stretch can also bias the final transformation. For example, if the textile is stretched in a uniform manner, it will shrink uniformly when released. If the textile is stretched more in one direction, however, it will undergo a greater shrinkage force in that direction when it's released. After stretching the textile, we add rigid or flexible material layers, like nylon, on top of it. This layer embeds the geometric information and pattern that will direct the precise transformation of the shape.

The type of material, the thickness of the layer, and the two-dimensional or three-dimensional pattern placed onto the textile all influence how it behaves next. If the material is rigid or thick, the layer will likely have a greater force than the shrinkage of the stretch textile, and it will significantly constrain the material from transforming. In order to take advantage of the stretch textile force, the deposited layer can be flexible or thin in certain areas to add flexibility and allow for the three-dimensional transformation.

The two-dimensional and three-dimensional shape of the deposited layer also influences the pattern of transformation. For example, if you deposit a circle onto the textile and stretch the textile in a uniform way, then when you release it from the rigid plate, it will jump into a saddlelike shape called a hyperbolic surface. The pattern of the material can be used as a geometric code to promote complex surface transformations. This triggers the precise self-forming process, transforming an ordinary flat sheet of textile into a useful shape.

We recently applied this technique to developing active shoes. Traditional shoe manufacturing is an example of an industry that traditionally produces static objects by manually assembling different parts: the uppers, insoles, outsoles, and other components. If we consider just the uppers, for example, there are usually quite a few components, such as the vamp, the outside quarter, the inside

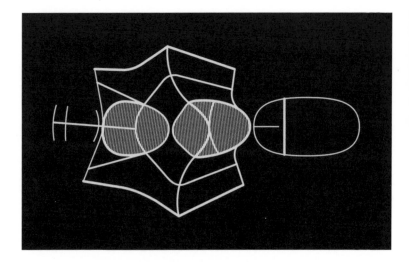

A two-dimensional pattern printed with a flexible polymer onto a stretch textile. The pattern encodes the geometric information to jump into the shape of a shoe when the textile is released after printing. *Credit:* Christophe Guberan and Carlo Clopath, and Self-Assembly Lab, MIT

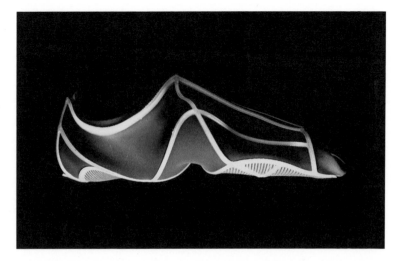

The final three-dimensional shoe that was created from a printed two-dimensional pattern on a stretch textile. After printing, the stretch textile jumps into the shape of the shoe. *Credit:* Christophe Guberan and Carlo Clopath, and Self-Assembly Lab, MIT

quarter, the strap, and more. Each of these components requires a significant amount of manual labor to assemble. The components need to be die-cut or laser-cut from leather or other materials. This is one of the most complex and labor-intensive aspects throughout the entire process. If the shoe is going to be made out of leather, the parts need to be arranged on a piece of leather, keeping in mind that the right and left shoe need to go together. Natural leather has a different amount of stretch across the different regions of the piece, which means that the various components of the shoe require a skilled hand for precise placement in order to meet all of the stretch requirements.

After carefully cutting the components, someone needs to form, sew, glue, and assemble them. This process can take many people, many machines, and many minutes, depending on the complexity of the material and the shoe. The manufacturing of shoes is still today mostly a manual process even for the largest and most technologically sophisticated companies. Similarly, the design process is traditionally separate from the fabrication and manufacturing process. Even in the most recent advances of 3D industrial knitting for shoe uppers, like Nike's Flyknits, the final product is designed to be static, it is not customized to the user, and it requires manual bonding or assembly for the sole of the shoe. Only recently have design and manufacturing started to inform one another and blur the lines between conception and creation with active materials.

The process of manually forming shoes is precisely what we targeted in a collaboration between our lab and product designers Christophe Guberan and Carlo Clopath. We wanted to see how we could simplify the assembly process by taking advantage of material transformations. In order to design an actively self-forming shoe, we had to identify the "geometric code" that we would print onto the stretched textile that would allow it to transform into a shoe.[6] First, we stretched the elastic textile around a rigid plate in a uniform manner. We then printed a polymer onto the stretched textile in a specific pattern. The textile was stretched in a uniform manner and the material properties were kept constant while the design variable that we adjusted was the printed pattern. The printing process allows for custom patterns and complete control over the shape while testing it out; once a shape is defined, it can be laminated, bonded, sewn, or otherwise combined with the textile.

We designed the printed pattern to create all of the curvature of today's shoes with a single piece of textile wrapping the foot from the toe to the heel. We went through many iterations and tested patterns to promote the precise transformation. Ultimately, we identified a pattern, printed onto the textile, and released it from the plate. The textile instantly jumps into its three-dimensional shape, encoded with the shoe's curvature to self-form into a footlike shape. As an extension of this process, we also created the sole of the shoe by promoting further curvature and wrapping from the bottom up around the sides. We didn't produce this shoe to have it manufactured and sold—we developed this as an experiment to see if it was possible and to challenge today's product manufacturing. We were conceptually pushing for an alternative approach to manufacturing that takes advantage of highly active materials. Much like a concept car, this shoe shows us what is possible and hopes to change the mindset of consumers and brands to create more active products in the near future.

Recently, we wanted to look at the adaptability of our textiles to address changing functionality or comfort requirements, while the product is in use. We wanted to go beyond just the shape-change of a textile and create porosity change with new functionality built directly into

A knit textile sleeve that senses and responds to moisture and pH to transform physically and visually. The moisture causes the knit textile to contract while the fibers turn pink when they sense a change in pH. *Credit:* Little Devices Lab, MIT, Self-Assembly Lab, MIT, Ministry of Supply, University of Maine, Iowa State University

A knit textile garment that senses and transforms based on temperature to adapt to the person's body. This textile structure can be used to create custom-tailored garments without expensive and labor-intensive cut-and-sew methods. *Credit:* Self-Assembly Lab, MIT, Ministry of Supply, University of Maine, Iowa State University

the textile from the filaments, fibers, and yarns all the way up to the garment. To accomplish this we worked in collaboration with Ministry of Supply and other researchers on a project through an organization called Advanced Functional Fabrics of America (AFFOA). The first development was focused on a single-direction transformation, where the textile could transform only once and never again. This type of single-direction transformation was geared toward tailoring and creating customized products that fit an individual's body. Typically, tailoring is only possible either by manufacturing a custom garment, which is often logistically complicated, expensive, and slow, or by manually cutting and sewing in a traditional tailoring process, which is often labor intensive and expensive. For these two reasons, we are used to seeing standard sizes like small, medium, large, and extra-large in mass-produced garments that don't fit perfectly. Similarly, even the same product and same size can be completely different depending on what factory it came from. With our research, we showed that we could still mass produce garments, taking advantage of the speed, scale, and efficiency of industrial knit textile manufacturing, yet we could activate garments to self-transform around the customer.

There are a number of examples where companies are trying to mass customize textile products using industrial knitting, either flatbed knitting or circular knitting. The dream is that you can go from a 3D body scan of the customer, and then directly manufacture a unique garment and ship it to the customer's door. This is extremely challenging logistically, however, because the custom program to run the knitting machine is not automated and the lack of dimensional precision in textile manufacturing makes this an unsolved problem. Our approach was to avoid the custom program and custom manufacturing challenges and focus on embedding the customization intelligence directly into the textile, not the machine. This allows us to mass produce standard sized garments, but when they arrive at the store, the garment can be activated with heat or moisture and it will then transform itself, adjusting directly to the customer's body. In this way, the customer receives a uniquely tailored garment that fits perfectly, without the complexity and cost of custom manufacturing or cut-and-sew tailoring. This type of single-direction transformation will happen only once; it will not return to the original shape and won't transform accidentally when the customer wears or washes it. It is only designed to adapt for their perfect fit.

In the more recent developments of this research, however, we have been able to demonstrate reversible, bidirectional transformations of textiles that are designed more for climate adaptability, allowing the textile to transform based on fluctuations in the external environment or the customer's body temperature. These also utilize industrial knitting technologies where we can swap the fiber, filament, or yarns, on a stitch-by-stitch basis, across the entire garment. That means, much like multimaterial 3D printing, we can change materials at every "pixel" (which in this case is a stitch in the textile) in a three-dimensional garment. By changing the materials in relation to one another, we can finely tune the various material properties, designing them to expand or contract based on external temperature or moisture change. Natural materials like wool or various polymer fibers will shrink with a certain temperature or moisture activation. We can then vary the knit structure, stitch by stitch, across the garment to change the way that the textile will move. A contracting zone can pull open certain pores, or lift a vent flap to adjust breathability. Some fibers will shrink, while others will bulk and expand in cross-section. We can use these behaviors to transform the global shape of the garment, creating zones that are thicker or thinner for insulation, comfort, breathability, or better fit. With this development, we can create knit textile garments that adapt for thermal comfort: if someone walks from their warm and cozy house to the brisk, cold outdoors, their lightweight and breathable sweater can close its pores and get thicker to help insulate and warm their body. Or vice versa, if they are in a cool air-conditioned office and they walk outside on a hot summer day, their garment should be able to open up and become more breathable and thinner, more lightweight, to help cool their body and stay comfortable in both temperature extremes. This type of active, self-transforming textile garment is bidirectional and can continuously adapt, going back and forth, adjusting to the ever-changing temperature dynamics that we experience every day.

These are just a few examples of active products from the Self-Assembly Lab's work. But this approach is not only happening in our lab; there are various other researchers developing unique approaches to the design and fabrication of these dynamic systems. For example, Hiroshi Ishii's Tangible Media Group at MIT's Media Lab has developed a number of active and self-transforming systems geared toward creating new interfaces between humans and computers that go beyond the keyboard or the screen.[7] They have developed inflatable devices, haptic interfaces, transformable surfaces, and a number of other technologies. Beyond these interfaces, one of their projects, bioLogic, has created a biological material actuator and sensor that can be applied to textiles to create active garments and other products. They created a shirt that can sense sweat and transform itself to open pores and keep the body cool. Similarly, Fiorenzo Omenetto's Silk Lab at Tufts University is at the forefront of biomaterial-based transformations, creating everything from new wearable sensors, to optically changing and transmitting materials, implantable electronics and medical devices, even dissolvable devices, all out of silk.[8] One of the mantras of the Silk Lab's work is "programmable form + programmable function = unique material outcomes," which succinctly articulates the ethos of programmable materials. This equation for programmabil-

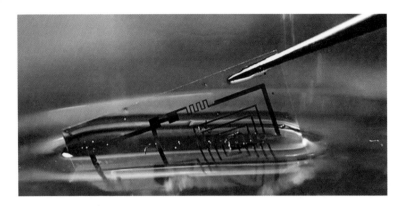

An electronic device made from silk that can be dissolved in a solvent, going from a functional circuit into a functionless liquid for recyclability or disintegration inside the body. *Credit:* Fiorenzo Omenetto

ity can be created both with biomaterials, as shown in these researchers' work, as well as synthetically created with non-biomaterials like films, textiles, and other material products. These projects, like the previous examples of self-forming footwear and active textiles, offer a new perspective on the agency of our materials, arguing for a more dynamic and ever-changing performance relationship with our products. While some of these products currently exist only in the lab, not on the market, it's likely not because they are more difficult to produce or more expensive or less durable. Their scope is limited today mainly because manufacturers and consumer cultures haven't yet made room for thinking about active products in this new way. But that will eventually change.

These materials, unlike the static products of our everyday world, do not resist all forces; instead, they become highly active, take advantage of the forces around them, and make use of their inherent material properties. Products shouldn't sit around passively—they should adapt to our needs, react to the environment, and push us to perform better and live healthier lives together.

Robots
without Robots

MOST OF US MAY STILL THINK of robots as mechanical and electrical devices that sense, respond, and move around. We might recall the types of robots we played with as children or that we've seen in movies, or maybe even industrial robots such as those prevalent in the automotive industry. All of these robots operate using computer programs: they compute information using traditional electronics, and move with actuators. Since objects can't move without something compelling them to do so, these robots tend to confront and fight the forces around their environment: walking robots try to overcome gravity to climb stairs or walk up hills; industrial manufacturing robots will try to fend off moisture, vibration, or any other external forces that degrade precision, maintenance, or efficiency. The forces a robot contends with are often seen as a nuisance, and the aim is to resist or fight these forces—vibration, moisture, temperature, sun exposure, and the like.

But this doesn't have to be the case. Rather than fight the surrounding forces, we can incorporate materials in such a way that we can take advantage of this energy to do useful work. Harnessing the very forces that have long been considered obstacles to overcome in the engineering process can radically improve a product and its performance. Consider, for example, our approach to flight.

A plane wing is a very specific type of electromechanical, hydraulic, or pneumatic system that transforms in quite dramatic ways. It's designed to move its flaps, create lift or drag, open vents, and produce the mechanical maneuvers that are required of an airplane in flight. All of the electronics, actuation, mechanical systems, structure, fasteners, and skin need to resist extreme temperature, pressure, moisture, and vibrational forces. These forces are quite significant, occurring regularly, every day, and they stem from the environment—flying from 30,000 feet to the ground and in one extreme weather condition to another; performing from 500 mph to a full stop and from low altitude to high altitudes and back. It's a wonder that planes fly so consistently and safely. These machines are designed to tirelessly fight every one of these forces, and thankfully they do a great job. But must they resist rather than harness them? Compare this to how a bird flies, actively working with the forces surrounding it to create flight by soaring with rising air pressure or adapting to wind direction. In a plane, the mechanical flaps and engine do all of the work, and the rest of the structure is designed to be static.

All of the mechanisms of today's popular robotic transformers are unlike the natural transformers that surround us every day. Humans and animals grow, repair, and adapt to their environments. Even nonliving systems can change their state from a solid to a liquid to a gas and back, with the right conditions. The material world around us is simple yet highly active, and perhaps far more advanced than the world of mechanical robots.

Harnessing these materials and forces means that robots do not always need robotic mechanisms to function. Robotic systems can get rid of the traditional motors, sensors, or electronics yet still have all of the lifelike qualities that we're used to seeing in the natural world around us. The future of robotics is drastically changing to encompass soft-material systems that look nothing like today's electromechanical robots yet have all of the existing capabilities and many more. This new breed of soft-material robots tends to be cheaper and easier to produce; they don't rely on batteries or electricity; they are disposable, recyclable, and extremely safe. It's tempting to call this soft robot something else entirely—after all, it's just *material that behaves like a robot*.

4D Printing

Early in my career at MIT, I worked on a research project at the Center for Bits and Atoms (CBA), as part of a Defense Advanced Research Projects Agency (DARPA) program called Programmable Matter.[1] At the time, programmable matter often meant electromechanical robots. Many of the research groups under this program were developing reconfigurable robots of various sizes, shapes, and functions: robots that could move themselves around, assemble or disassemble through modular components. Our project created a suite of one-dimensional reconfigurable robot chains that could self-transform into any 2D or 3D structures. Like robotic worms, these structures could squirm around and fold or curl themselves into different configurations. These robots ranged from roughly a centimeter in size all the way up to meters. I worked on the largest of these robots, at the meter scale.

Later, as I looked back on the project from a broader perspective, I started to reflect on some of the limitations of building with electromechanical robots. As an architect, if you hypothetically make every brick out of a robot, then the building's construction will quickly become too expensive due to the cost of the components and the power required to operate. The performance of

A large-scale reconfigurable robot developed under the DARPA Programmable Matter program. The robot could transform from a one-dimensional strand into a two-dimensional pattern and a three-dimensional structure. *Credit:* Skylar Tibbits and Center for Bits and Atoms, MIT

the robotic bricks would likely fail too often because of the number of components and it would be more complex to assemble compared with a traditional nonrobot construction process. This shows some of the scalability concerns when attempting to build with robotic components either at large scale or with large quantities. With this in mind, I was trying to figure out ways to create the reconfigurable robotic behaviors without robotic components. One of the Self-Assembly Lab's first breakthroughs was employing multimaterial printing to create reconfigurable structures without robotics. We called this work 4D printing.

Multimaterial 3D printing is a process of depositing different materials simultaneously in three dimensions. We took this one step further by printing with different material properties to enable the parts to transform from one shape into another. We called this 4D printing because we wanted to print 3D structures that would transform or reconfigure over time. One of the materials we printed was a hydrogel, which can expand based on the amount of moisture in the environment. The other material was a rigid polymer, which we could print into

4D printing is a method of producing multimaterial structures that can sense and transform based on moisture activation. Materials with differing properties are printed together, one of them swelling when subjected to moisture, creating a physical transformation based on the precise geometry of the second material. In this method, flat sheets can be produced that transform into precise three-dimensional objects. *Credit:* Self-Assembly Lab, MIT, Autodesk, Stratasys

precise joints that would encode (through geometry, as shown in the previous chapter) all of the details and structure needed to transform from one shape to any other. We printed different types of structures, which were either flat surfaces or 1D lines, to minimize print time and the amount of material, and then we allowed them to self-transform into larger 3D structures when they were placed underwater. Single lines transformed into macroscale proteins; lines folded into the letters "MIT"; surfaces folded into 3D cubes; surfaces created curved-crease origami; structures locally expanded or contracted from a flat sheet into a doubly curved mathematical surface; and so on. All of these models demonstrated the increasing complexity and physical ability of our 4D printed structures to transform from any one shape to any other without robotic or electronic assistance. In my mind, these were versions of the robots that we had made under the DARPA program, but now I could easily print them—there was no assembly required to make them work, they weren't expensive to produce, and they didn't fail as often as other robotic systems.

We weren't just 3D printing static *objects* that sit on your desk, or trying to fabricate something in a new way, as we had seen with other 3D printed projects. Rather, we were interested in printing highly active, self-transforming *materials* that would adapt and reconfigure over time. In many ways, this work was similar to the electromechanical robots we had developed previously, but we did not have to assemble all of the robotic components. It was also similar to printing traditional off-the-shelf "smart materials"—like shape memory alloys or shape memory polymers that change shape when subject to temperature or electricity—but we could now create them from scratch and customize their material properties, enabling unique transformations tuned to their environment. Today's smart materials are often limited by fixed shapes,

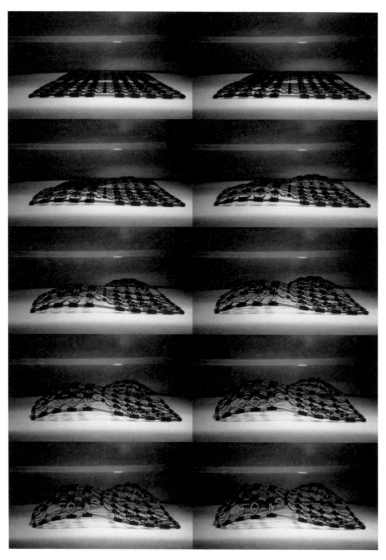

A 4D-printed object that transforms from a flat sheet into a complex surface without any manual forming or manipulation. The printed structure includes materials with differing properties, one of them swelling when subjected to moisture, creating the undulating geometry by both curling and expanding across the surface. *Credit:* Self-Assembly Lab, MIT, Autodesk, Stratasys

sizes, and material properties, are often more expensive than traditional materials, and require extra assembly to embed them in products—which in large measure explains why they aren't more widely used. Our 4D printed smart materials could be customized to fold, curl, stretch, or shrink based on water, temperature, or light.

Since the Self-Assembly Lab first introduced 4D printing in 2013, a growing field of researchers has emerged, with journals, conferences, and even grants dedicated to the topic.[2] One of the most surprising yet promising developments has been within the medical field for adaptive surgical devices. Beyond traditional electro-mechanical robots and static medical devices, applying 4D printing to develop customized surgical devices to solve pressing medical challenges is one of the most important applications of this emerging technology. The devices can be designed, customized to the patient, printed, and then inserted into the body to adapt and transform in precise ways. These devices can solve a localized problem within the body and not require human or external robotic surgery. This research is already happening in a number of places around the world from drug delivery capsules and stents to airway splints.[3] For example, doctors in Michigan have 4D printed and then implanted airway splints in three infants who had a rare life-threatening disease. These devices were designed to grow and transform inside the body and then dissolve over time, eliminating the need for invasive surgery to remove them. This is likely the next generation of smart medical "robots," made of simple materials and easily fabricated with customized behaviors.

Using technologies like these, we can potentially see a future when many products are printed with all of the intelligence embedded directly into the material, not requiring additional assembly or components to make them "smart." As I look back to the DARPA program on

Programmable Matter, perhaps one of the reasons it may not have been quite achievable a decade ago was the limitations and lack of scalability when working with electromechanical robotic systems. Many of these researchers have since shifted to using soft, agile, flexible, adaptable material robotics. From granular jamming to synthetic biology, soft robotics, 4D printing, and a great deal more, researchers are pioneering a new wave of materials and fabrication processes to finally realize this vision of programmable materials.

Material Actuators and Sensors

Materials *are* robots. Materials can exhibit actuation and sensing, and contain embedded information without the need for any external devices. This capability can be seen from the smallest of scales to the largest, from self-folding DNA origami that folds from a single line into an arbitrary shape, to large pneumatic exoskeletons that transform with speed, precision, and flexibility.

Take wood, for example. Wood is quite possibly one of the "smartest" natural materials we can find. Wood behaves as both a sensor and an actuator: it both senses the moisture in the environment and then physically transforms from one state to another. It also encompasses geometric information in its grain pattern. Recall the woodworker taking advantage of the swelling behavior of wood to create precise and extremely strong joints. The wood's self-actuation comes from its cellulose-based composition, which senses and absorbs moisture, causing the wood to swell. When there is a certain percentage of moisture in the environment, the wood will curl into a predetermined shape. Without moisture, the wood will remain static and flat. This capability makes the wood both a sensor and an actuator with the instructions embedded within its grain.

Material sensors appear in other places as well. If you have a typical thermostat on the wall of your home (not a "smart" thermostat), it likely uses a bimetallic strip to sense the environment and create physical changes to rotate a needle indicating the current temperature. Just like Harrison's sea clock of the 1700s, this contemporary material sensor is made out of two metal layers, and each one has a different coefficient of thermal expansion—which is the amount that a material will expand or contract with a temperature change. When two metals with different coefficients are bonded together and one of the metals expands or contracts, the other one is forced to curl. This curling effect is exploited in the thermostat to rotate a needle. The bimetallic piece is extremely sensitive, which makes it a fairly accurate method of measuring temperature. The piece will last for decades if not centuries, since it doesn't require motors, external power, or other components that could fail or otherwise degrade. In fact, the material sensor will likely far outlast its electronic counterparts. This material sensor and actuator is cheap, robust, simple, and extremely smart, even if it's not called a "smart" thermostat.

Material Logic

Materials can also embody logic. One of the simplest examples of this is at the top of a ballpoint pen. Some pens have a button that you can press to extend or retract the tip. This button takes a single input, a downward press, and performs one of two possible outputs. The behavior of the output depends on the previous state, which is an example of hysteresis. If the pen is extended, then it will retract the next time you push the button; if it's retracted, then it will extend the next time, and so on. We take this simple system for granted. But it's fascinating, as it demonstrates an important point in material logic:

```
if button is pressed:
        if pen is extended:
                retract pen
        if pen is retracted:
                extend pen
else:
        keep the current pen position
```

This extremely simple material device actually stores memory. It stores the previous state, responds to an input, and makes a logical decision to either change state or not based on a previous condition. It demonstrates a mechanism of logic embedded in the structure that requires no electronics, transistors, or typical computing devices.

If we apply this new way of thinking about material sensors, actuators, and logic to the airplane we imagined earlier, we can envisage a radically different future for aviation. Could a soft, morphing plane component with *birdlike* efficiency be possible? A recent collaboration

A new method of creating flexible carbon fiber structures that can sense and respond to temperature change. Flexible carbon fiber sheets are printed or laminated with polymer layers in precise geometric patterns. When the polymer layers sense environmental changes, they promote the three-dimensional transformation of the carbon fiber. *Credit:* Self-Assembly Lab, Christophe Guberan, Erik Demaine, Carbitex LLC

between our Lab and Airbus brought this idea one step closer to reality.[4] We were focused on the air inlet, a small component at the top of the jet engine that brings in air to cool the engine. Unfortunately, it also causes drag and reduces the efficiency of the plane since it is essentially a hole in the top of the engine.

The conventional solution for this type of problem in aviation is to create an electromechanical, or hydraulic flap that opens and closes in the same way that a plane's wing functions. This solution requires additional motors, sensors, electronics, and wires that run back to the cockpit for control, all of which impedes the plane's manufacture. The additional parts also add to the overall cost to build the plane, require more maintenance, have more steps in the assembly process, and increase the likelihood that a part will fail. All of these challenges with the robotic approach could lead to a solution that was potentially less efficient than not having solved the problem in the first place—therefore increasing the cost to fly because of the additional components, maintenance, and fuel. In light of this, Airbus challenged us to find a unique solution that could control the airflow without adding levels of complexity.

At the same time, we had been developing a new active material composition in collaboration with Carbitex LLC, a company that makes flexible carbon fiber material. We wanted to extend the material functionality, and so we developed a way to promote carbon fiber to self-transform based on temperature, moisture, light, or pressure. To do this, we combined different polymer layers with the carbon fiber. The orientation of the polymer layers in relationship to the orientation of the grain of the woven carbon fiber would dictate the type of behavior—if the carbon would fold, curl, or twist, for example. The thickness of the polymer layer would dictate the amount of force, the speed, and the amount of total movement. The

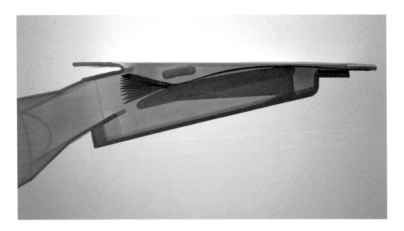

A carbon fiber component placed within an air inlet component for the Airbus engine. The carbon fiber piece opens and closes the hole to control the amount of airflow into the engine. This movement is created by either temperature change or pressure differential rather than electromechanical actuation. *Credit:* Self-Assembly Lab, Carbitex LLC, and Airbus

type of polymer influenced the activation (temperature, moisture, or light), so we could select different combinations and then combine them in specific geometries in relation to the woven pattern. This active carbon fiber construction was a perfect solution to the Airbus conundrum because it was extremely light and strong like traditional carbon fiber, which the aviation industry uses on a regular basis. However, now the material could be highly active, sensing the right conditions, making the decision when to transform to the desired state to bring cool air to the engine, and ultimately returning to the closed position.

Ultimately, we developed a flap that transformed based on a pressure differential from the inside of the engine to the outside. The flap remains open when the plane is on the ground in order to maximize the amount of airflow to the engine. Then, as the plane takes off and starts to gain altitude and wind speed, the flap jumps into a secondary state and closes to reduce the airflow to the engine. This secondary state is created with a bistable mechanism (a mechanism that is able to rest comfortably in two different positions) and induced by the pressure differential between the inside and outside of the component. As the plane returns to ground, the flap closes again. We sent the component through wind tunnel testing at Airbus's facility in France, successfully demonstrating the functionality of the morphable air inlet component. This design didn't rely on the electromechanical and robotic devices that currently operate our planes. It can sense and actuate with encoded behaviors. This example demonstrates that robotic-like capabilities can be embedded in simple materials without additional components.

There have recently been a few other examples of morphing aviation components. A team at NASA and the Center for Bits and Atoms at MIT developed a wing made up of an assembly of composite structures that can morph and twist to control its flight behavior.[5] The main goal was to demonstrate morphability to go beyond the mechanical flaps and devices, embedding this into the entire structure and flexibility of the wing. This example and various other morphing developments for aviation components point to a new model for increasing efficiency in flight, reducing assembly time, reducing complexity, and minimizing maintenance. All of these technological advances are made by simply using less material in a smarter way.

Printable Robots

There is a growing number of robots that may on the surface look like classical robots but do not have any of the usual robot organs. Researchers are developing "printed robots" and self-folding or self-assembling material robots.[6] These simple yet smart structures are good examples of the new class of robots that are easily produced—often printed, laser-cut, or otherwise digitally fabricated—and are high-functioning with respect to cost, durability, or strength-to-weight ratios. Some of these robots self-transform from fabricated flat material sheets into fully functioning forms that can walk across a surface, or pop into a robot and then fly off the table, or an inflatable silicone structure can bend and conform around an object to make a gripper. For example, Daniela Rus's Distributed Robotics Lab at MIT revealed a millimeter-size piece of material that self-folds and can walk, swim, and even dissolve itself for recycling.[7]

More recently, George Whitesides, Jennifer Lewis, Rob Wood, and their respective teams collaboratively developed a multimaterial, printed, chemically actuated, fully autonomous soft octopus called the Octobot.[8] This robot does not require a tether to electronic or pneumatic

The Octobot is an autonomous, multimaterial printed soft robot with internal microfluidic logic that modulates the flow of fuel, creating a chemical reaction to propel its movement. *Credit:* Photo by Lori K. Sanders, Harvard University. Project by Michael Wehner, Ryan L Truby, Daniel J Fitzgerald, Bobak Mosadegh, George M Whitesides, Jennifer A Lewis, Robert J Wood

control; rather it uses microfluidic logic that modulates the flow of fuel, creating a chemical reaction that results in a gas expanding a portion of the robot, propelling it to move. In all of these examples, the robots aren't made with clunky metal components, gears, or traditional actuators; they are made from sophisticated material composites through novel fabrication methods that enable their programmability.

Needless to say, this new class of smart material structures is challenging the way we think about producing robots. This growing field of soft robotics often produces material structures that are lighter, stronger, faster, more adaptive, and more durable than traditional electromechanical robots. Programmable materials are allowing us to completely rethink *robotic performance* as distinct from traditional robotic capabilities and construction. But even more important here, these material robots invite new applications that haven't used robots before because they were too expensive, too dangerous, or too complex to assemble. For example, Rob Wood and his team at Harvard developed a soft robot in collaboration with David Gruber and National Geographic, which was used as an underwater gripper on remotely operated vehicles to gently grab and investigate marine life.[9] Soft, autonomous robots could also be used in dangerous situations like war zones or for investigating collapsed buildings or hazardous sites, and in various other scenarios where it would be too difficult or dangerous for people to explore. Given their flexibility, they can adapt to different environments, squeeze through tight spaces, and flex to accommodate different forces, which would be difficult for traditional robots. Soft robots have also been used as interfaces for the body in prosthetic, exoskeleton, or haptic feedback devices because they are soft and responsive and tend to be safer for human contact than traditional mechanical robots.[10]

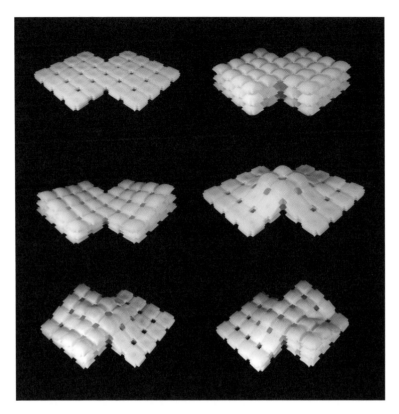

A printed inflatable material that transform and morphs from one state to another based on air pressure. The printed structure can actuate vertically, undulate, or create a wide range of configurations. The pneumatic control of the soft silicone structure points toward a future of cushions or foams that can change stiffness, adapt to the form of the body, create tunable comfort, and provide lumbar support, massage features, or even crash protection. *Credit:* Self-Assembly Lab, MIT, and BMW

Smart Wearables

As discussed in previous chapters, the apparel and foot-wear industries are poised for change because of soft-material robotics. Companies in this space have long desired ways to incorporate robotic-like capabilities in the design of their products. Since at least the days of the self-lacing shoe featured in *Back to the Future*, or Ironman's smart exoskeleton, many companies have dreamed of smarter products that can morph to be one with the body, adjusting their shape for comfort, changing their stiffness or traction for increased performance, and tuning their porosity for breathability and weatherproofing. Often, these companies see the most direct pathway to realizing these capabilities is to add motors, electronics, sensors, and other robotic-digital components to their products.

In a similar vein, we are witnessing a boom in smart "wearables" and sports technologies: from wristbands that track our every movement and heartbeat, to shirts that monitor your vital signs, to clothing that actively heats or cools. It seems as though many brands are chasing after smarter and smarter wearable garments and gadgets. The simple addition of robotic components makes the product more expensive to produce both in terms of the material components as well as the additional time and complexity of the manufacturing and assembly processes, costs that are usually transferred to the customer in higher prices. Robotic components also add weight to the garments: heavier and bulkier shoes and clothing are rarely ideal. The manufacturer has also increased the potential for failure; the garments often require power and batteries, and, worst of all, the chances that they'll injure someone goes up, as they may overheat or break. We may want smarter products, but most of us simply don't want robots in our clothing.

Soft, adapting, material robots can provide us with all of the same features for smart wearable clothing, shoes, exoskeletons, and any other product that comes into contact with the body. They can operate off of body heat, sunlight, moisture, or the abundant natural energy sources that surround us, rather than batteries or wires. They can sense our body, transform in soft, squishy, comfy ways, and be seamlessly integrated into our everyday products. These material robots can even be produced using textiles and other sheet goods with existing manufacturing processes like weaving, knitting, and lamination.

We should challenge ourselves to create more intelligent, yet simple, material robots. We should measure programmability and functionality against the complexity of the robot. As functionality increases, complexity, cost, and failure tend to increase as well. Through programmable materials, however, the complexity of the robot can decrease while its functionality and capabilities increase. This is true elegance—elegance both in functional value and as a technological feature.

Computer scientists often talk about "seeking elegant code"—in other words, they want to find the most efficient and simplest solution to a complex problem. How can you reduce the number of lines of code to accomplish the maximum goal?[11] Physicists and mathematicians, too, are always searching for the simplest explanation for the most complicated questions. We need to translate this goal of minimalism to robotics. Achieving an elegant and minimal solution requires a process of refinement and iteration. We should go beyond the complex solutions by enabling materials to achieve maximal elegance through their own evolution and adaptation.

Build from the Bottom Up

HUMANS TYPICALLY BUILD from the top down. We come up with ideas, we secure the materials, and we put them together to realize those ideas. Creativity comes from us, the physical reality comes from the assembly of materials by builders or robots, and functionality emerges from the devices. In this view of the world, materials are passive and they listen to our commands.

But there is another way to consider design, functionality, and assembly, one that emerges from the bottom up. This approach enables materials to interact with one another and their surrounding environment so we can create systems that amount to far more than the sum of the parts. Look at the complexity, functionality, and design of a human, a plant, a diamond, a mountain, or a planet. All of these natural systems, some organic and some inorganic, are formed from the interaction between the components and their environments, rather than top down from an a priori design. There aren't little construction workers with tiny sledgehammers and screwdrivers making our cells or sculpting a mountain. There are no planetary-scale printers making planets. Design, construction, and functionality emerge on their own.

We can see this type of bottom-up construction happening with many living species aside from humans. Ants, bees, beavers, birds, and nearly any animal that builds structures deploys a process of bottom-up assembly. Sometimes these species build individually, following rules of their own and adapting based on the forces and changes in their environment or materials. Beavers, for example, don't start with an initial design; rather they are constantly building and negotiating between the found materials and the forces of the environment to build a functional dam. Their material components are always changing based on what they can find or manipulate, and the forces of their environment are always changing as water levels or weather patterns fluctuate. The only way they can build such amazing structures is by adapting on the fly in collaboration with materials and their environment. Other species work together through a form of collective construction, called stigmergy, where agents work together, not through direct coordination but rather through indirect interactions with one another.[1] Often this is performed by an insect leaving a chemical trace in its environment that prompts another insect to behave in a particular way. Collectively, they can work together and build something without having global or even local instructions and without directly collaborating. They listen to the environment, work with the materials at hand and build incredibly complex structures like termite mounds, beehives or ant farms. Both of these types—bottom-up assembly by individuals and collective construction with many agents—are simple, yet powerful techniques that we could also employ in our human-built world.

Bottom-Up Construction

With hundreds or thousands of components in the assembly of any sophisticated manmade system (cars, planes, buildings, etc.), it can be quite difficult to explicitly prescribe the placement of every component to make an arbitrary design. This is especially complex given the whisper-down-the-lane challenges with communicating instructions to construction teams or if the placement of each piece is dependent on the placement of various other pieces. These types of interrelationships between components and construction teams can make assembly difficult in ever-changing environments, and thus it is sometimes valuable to utilize bottom-up assembly, where the structure emerges from the relationship of the components and their environment, not from predetermined protocols.

These bottom-up processes can also be used in the physical organization of patterns and structures. For example, a team of researchers at Harvard has shown that distributed robotic assemblers can place individual components and then climb those structures to continually build something larger.[2] Similarly, Radhika Nagpal's Kilobot swarm research created large numbers of simple, cheap robotic components that have decentralized communication and simple programs, yet collectively they can work together to assemble into precise structures, much like ants or schools of fish, forming two-dimensional images, symbols, or letters.[3] Amazingly, this assembly process happens without centralized control.

At larger scales, research teams like Gramazio Kohler Research at ETH Zurich have created drone-assembly techniques, where flying drones can pick up simple bricks and stack them to build large-scale architectural towers. Or similarly, the drones have built tensile bridge structures spanning from various points, with configurations completely made out of ropes tied together by drones.[4] The drones need to coordinate their movements and placement with one another, working with relatively simple components to build collectively. This specific example is closer to top-down design with bottom-up construction in that the drones execute a predetermined design. Even though the overall design may not change, the drones must be able to adapt on the fly based on the

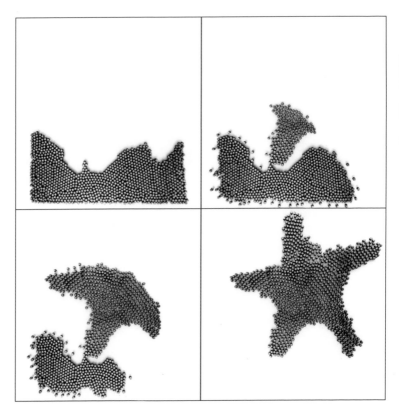

More than one thousand simple robotic components, called Kilobots, that have decentralized communication and simple programs can work together to assemble into precise structures, forming two-dimensional images, symbols, or letters. *Credit:* Michael Rubenstein and Radhika Nagpal, Harvard University

other drones and progress made throughout. This is useful in a construction process with many more "hands" to assemble something, making it faster, more distributed, and more agile in adapting to changes. However, these robots can also be autonomous or work in tandem with humans to be able to update the design and construction on the fly. This is bottom-up design with bottom-up construction. Gramazio Kohler Research has also developed capabilities for on-site robotic construction where

Flight Assembled Architecture is an architectural tower installation assembled by flying drones, without any human or top-down robotic assembly. *Credit:* Photos by François Lauginie and Gramazio and Kohler. Project by Gramazio and Kohler and Raffaello D`Andrea in cooperation with ETH Zurich

the robot adapts to changes made by the human or the environment. By continually scanning the placement of the component and structure, the robot can adapt if a component is misplaced, or adjust if errors begin to accumulate or there are structural failures.[5] This points to a model of bottom-up construction and bottom-up design with tightly coupled human-robot collaborations.

Marcelo Coelho and David Benjamin developed a project that went in the opposite direction with smart components and simple human builders.[6] In their case, the bricks contained all of the information to assemble the structure, telling the people when and where to place the components. When someone picked up a component, it would light up and then the corresponding place in the overall structure would also light up the same color to match. This color-coded lighting pattern informed the user where to place the part in the accurate location. The collective information embedded in the bricks ensured that even the uninformed human assembler could precisely build the desired final structure.

Coelho developed an even more low-tech yet vastly larger version of this system for the Rio 2016 Paralympics opening ceremonies.[7] With simple light sticks that contained long strips of LEDs that would flash and fade with digital light patterns, hundreds of people were able to be guided through choreographed patterns of physical movement at the scale of a stadium. The light sticks displayed patterns that were unique to the individual, their timing, and their relationship to other individuals, and in this way a swarmlike, bottom-up process was created with humans and simple materials. The system was surprisingly simple in the amount of information that each person knew or the information in the components, yet it created beautiful, almost hypnotizing, patterns. Perhaps some of these would have been possible with top-down instructions, more like a choreography with someone teaching the moves and training for a specific pattern of movement. If they wanted to change the pattern or have a different number of people, the top-down approach falls apart because of the drastic changes, human errors, and less than perfect memory. In this example, the power of collective, bottom-up instruction was given by a simple set of rules embedded in the fundamental building block, the light stick, and collective organization, precision, and elegance emerged.

If we translate this to the construction of our built environment, we could develop a new approach for building faster, cheaper, or in extreme circumstances. This could enable building precise structures without precise components. Or perhaps construction crews could have unskilled workers, people with very little assembly information, who could build precise things in collaboration with smarter material components. One of the challenges of construction is the assembly sequence, where one task cannot start until another task is finished. With collaborative bottom-up teams, humans, materials, and

2016 Summer Paralympics opening ceremony, which included hundreds of performers and simple LED light sticks that guided their movements through algorithmic patterns, generated from the bottom up. The light sticks displayed patterns that were unique to the individuals, the timing, and their relationship to one another, creating swarmlike behaviors that weren't taught or pretrained. *Credit:* Tomaz Silva / Agência Brasil

robots could adapt sequences on the fly, adjusting to one another or assembling many things simultaneously. The building process could also adapt throughout construction sequences without change orders and complex redesigns where different trades get out of sync with one another. The materials and builders could collectively construct through stigmergy by taking cues from each other and their environment.

Control

So why is it that we don't often see these principles used in everyday construction and manufacturing? It's sometimes difficult for designers to understand bottom-up processes, or to let go of agency and allow solutions to emerge over time. In some ways, bottom-up processes seem antithetical to the traditional notion of "design" where we, as humans and creatives, come up with an idea and make it happen. But this doesn't need to be the case. From a manufacturing and construction perspective, it can also be seen as risky to let go of control over the assembly process, because, in the end, the construction team is responsible for the final build-up. It goes without saying that it is important to design with safety standards, precision, and best practices of the construction or manufacturing industries. The two aren't mutually exclusive, however. We've been trained to think in a top-down style, one that ignores valuable solutions that may develop from the materials and their relationship to the surrounding environment.

We are often afraid of losing control, associating a lack of control with messiness, randomness, a lack of precision, or destruction. But we should recognize that we rarely have control—in the design process, assembly sequence, or even the final product, given the changing needs and environment of the real world. Buildings, for instance, are designed for one purpose and then repurposed (or torn down) as years pass, and it's impossible for architects to fully design for these future uses. In the assembly process, things change, whether those are logistics and supply-chain issues, or design changes, or environmental changes in a factory or a job site. Even when we have designed and built something with seemingly complete control, the final product can always be repurposed, hacked, changed, and misused. So it is perhaps in our best interest to strive for a more elegant balance and relationship among the design, the materials, and their intended environment.

Surrendering a bit of control can be productive, allowing us to tap into the intelligence in materials and the broader environment. Having full control can be limiting because we either have received information (this is how we have done it in the past, so we will do it like this in the future) or, in the best of scenarios, you have a singular idea. But there is an entire world of other possibilities.

For example, we have very little control over the growth of a tree, the assembly of a garden, the chemical reactions of yeast in brewing beer, or even the development of a baby (or any living organism, for that matter)—and yet incredible structures, design, and functionality emerge all the time, with minimal input from the outside. We cannot completely control the complexity of a living organism and the process of growth in the natural world. We can change the ingredients and influence the amount of energy we put into growing a plant or a human, but we can't change the cellular building blocks or the ways the organism interacts with its surroundings—those are built from the bottom up. The same holds true for countless other dynamic interactions in our daily lives.

Think of the weather. On a daily basis, we adapt to incredible changes in the weather. We have no control over the weather, yet we live our daily lives, adapt, adjust, and continue to be productive in many circumstances. We actually take advantage of this dynamism in the environment with different regions around the world that are famous for wine making, each specialized to their local environment and yearly weather fluctuations. If every location around the world had the same conditions, we would surely miss the diversity and quality of our wines. Unknowns can be challenging, but on the flip side of the coin, there is something gained—surprise, novelty, and complex differentiation in the delicate relationship with a process that isn't entirely in our control. Bottom-up design and assembly are proven processes in both natural and synthetic systems. Some of these processes demonstrate designs and outputs that are so sophisticated that they far outweigh what we can construct in other ways. By losing control, we can gain surprising new results.

I'm not proposing here that we use a random smattering of parts to build. The output of a bottom-up process does not have to be random at all. Even if we determine what the product will be beforehand—we may want to make a cell phone or a shoe or a chair—as we saw in previous chapters, the process of assembly can still be bottom up, and the parts can come together on their own. We can build in all of the mechanisms of self-assembly, from geometry to interactions and just the right amount of energy to make sure we assemble a strong product.

This could have particular advantages as things scale up. Some might argue that we could build a single product faster by doing it piece by piece with a human or a robot rather than through self-assembly. But once we need to build many things, manufacturing often scales in a linear fashion. You need more time, more people (more money), or more robots to build more things. But if things are assembling from the bottom up, they can be assembling in parallel, many pieces coming together simultaneously without more people or more robots. This type of process can quickly switch from building one product to building many products instantly. We can even adjust the type of products being assembled. This is very challenging for a traditional factory with people or robots that need to be rearranged, reprogrammed, and retrained. Fixed infrastructure and equipment are not as agile—scaling up or down or changing parameters—compared with a self-assembly approach. Parallel and bottom-up assembly becomes increasingly efficient compared to linear assembly as the number of parts scale or as the environments get more complex.

Going beyond top-down design with bottom-up construction, we can also allow design and construction to emerge simultaneously. At the Lab, we've worked on a series of projects that see the fabrication process as the design process and vice versa—that is, the design emerges as a structure is created.[8] In one experiment, we created a water-filled chamber where components can move freely, find each other, connect or disconnect, and build structures based on their geometry, their relationships to their neighbors, and their relationship to the environment. The forces in the environment have a direct impact on the structures that can emerge. If we create a vortex or swirling effect with pumps inside the tank of water, the units make a towerlike structure. Or if we have jets pushing the water upward, or buoyant units, then they form sheets at the surface of the water. The units, geometries, and connectivity to one another did not change; only the forces of the environment changed, which resulted in different potential design solutions. The environment, in other words, can influence the design and promote the emergence of successful structures, a sort of physical fitness criteria, where only the *best* design can succeed. Put another way, the design that succeeds may be the most *comfortable* or the easiest configuration for the structures to form given the forces of the environment. In this way, we can be open to surprising and unexpected solutions.

In another experiment, we created an aerial configuration of units that were subjected to wind in a vertical chamber, which caused them to tumble and fly around.[9] Much like kernels in a popcorn machine, the units moved around rapidly, bumping into one another. Due to the weak local force of the connections between the units, they easily connected and disconnected from each other. The units broke apart if they fell to the bottom (due to the force of hitting the ground) but they could connect again when they were pushed back up into the air. Over time, we found that certain configurations wouldn't fall and break— rather, they would hover until more units connected, which allowed them to stabilize and continue to fly. Flat surfaces or winglike geometries and other configurations emerged that were well suited for flying. This type of crude chamber promotes the simple *evolution* of flying geometries: the pieces will fall and break unless they form into a "flying shape." These may not be the most optimal designs

An air chamber with self-similar units that fly around due to wind turbulence. The units connect with one another, fall to the ground, and break apart. Only the most successful structures survive, evolving a fitness criterion for flight where the units hover in the air and fly in the chamber. *Credit:* Self-Assembly Lab, MIT

for flying, but they assembled themselves and found functional solutions that allowed them to fly without human input guiding their form. It is a simple example of design emerging from the forces of the environment—and how losing control can lead to benefits in functionality. The designs may reveal surprising and novel ways to fly. Or they may be intricate and complex designs that are so unique that we wouldn't have imagined them.

As another advantage of this approach, imagine trying to build something in a complex environment, whether that is underwater, outer space, war zones, after a disaster, or even within the human body. It is often difficult to predetermine the best solution and even more difficult to build when you can't actually see or access the site, as in surgical applications, or where conditions may be constantly changing. In these cases, it is more productive to design smarter material components that can be placed within the body and then adapt to the conditions of the environment, assemble themselves, and continue to transform their function to solve the local problems. Or imagine having to build in a location where you can't guarantee skilled labor, or where it is difficult to build traditionally. It would be advantageous for all of the information and agency to be within the material components, allowing them to work with people and assemble themselves, rather than try to predetermine the precision or skills of an unknown construction team. These are just a few examples where utilizing bottom-up assembly and smarter material components could offer significant advantages.

Destruction as Construction

The power of bottom-up approaches extends beyond the realms of product design and manufacturing, or building construction. A recent Self-Assembly Lab project in the Maldives has been exploring this potential at the environmental scale. We have been collaborating with a Maldivian team led by Sarah Dole and Hassan Maniku to design self-organizing sand systems that utilize wave energy attempting to grow islands and sandbars. Typically, ocean waves, storms, and currents are seen as destructive to islands and coastal regions, where they try to fight the forces of the ocean. In this project, we are using the natural forces of ocean waves and currents to create construction rather than destruction. This ambitious and radical project is attempting to overcome some of the challenges faced by low-lying island nations and coastal regions with sea level rise and increased storm inundation due to climate change. This project is extremely dynamic, with systems like weather patterns, fluid dynamics, sediment transport, bathymetry, and many other complex systems at play.

These complex conditions are a clear example of a problem that cannot be solved with traditional top-down design processes or a conventional relationship with materials and fabrication. The traditional attempts to solve this problem have been top-down and deterministic, and they have engineered solutions that often don't solve the problem, or sometimes even make it worse. The most common solution is to build barriers, walls, and other physical structures that fight the forces of nature. This approach takes the human over nature perspective, where we are the master designers and we will shape our environment, regardless of the forces of the ocean. We often think we can build in specific places because we are smarter, and nature will take a back seat. This method has proven to be unsuccessful in many instances, with levies breaking and seawalls inundated with water and sand. Sometimes these approaches may solve the problem locally but make it much worse somewhere else.[10]

The other common approach to this problem is to use dredging, pumping sand from the deep ocean back onto a beach or into a pile to build a new island from scratch.

A project in the Maldives designed to promote the self-organization of sand for growing new islands or helping rebuild coastlines. This approach uses wave energy and its interaction with submersible geometries to promote sand accumulation. The aim is to create a scalable and adaptable solution to help protect coastal communities from rising sea levels and continual erosion. *Credit:* Self-Assembly Lab, MIT, Invena, Taylor Perron, James Bramante, Andrew Ashton, Tencate, SASe Construction, Planet, Vulcan, Allen Coral Atlas

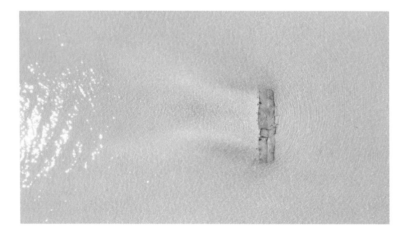

A submersible geometry deployed in the Maldives aiming to promote sand accumulation. This large-scale field experiment is 20 m × 4 m × 2 m and has generated an area of sand accumulation that stretches across more than 30 m × 20 m, creating hundreds of cubic meters of new sand in a targeted area. *Credit:* Self-Assembly Lab, MIT, Invena, Taylor Perron, James Bramante, Andrew Ashton, Tencate, SASe Construction, Planet, Vulcan, Allen Coral Atlas

This approach imagines that we know the solution to where the best island should be placed and how to restore the beach, but it ignores the forces at play and doesn't collaborate with the medium. Even worse, this method is really only a band-aid solution that may temporarily stop the erosion of beaches; year after the year, the sand will need to be redredged. Dredging is extremely energy-intensive, costly, and harmful to the marine environment, yet our top-down mentality continues to call for this type of approach.

Instead, we have taken what we see as the most elegant approach to this problem: to collaborate with the materials (sand) and the energy of the environment (waves, currents, storms, and so on). We can't possibly know the best final location or configuration of the sand. We can have a good idea as to the best approach, and we can have a desired location or end result. Then, we can collaborate with the materials and environment to promote certain scenarios to arise and let the force of the water guide the organization of the structures. We have now conducted multiple full-scale field experiments in the Maldives that we continue to analyze through satellite imagery, drone footage, and physical measurements over months and years. Simultaneously, we have been working on hundreds of lab experiments and simulations with a wave tank, changing the properties of the system with different wave patterns and different underwater geometries to promote sand accumulation in short periods of time. Through these lab experiments we then take the best of what we see, build them at large scales (~20 m × 4 m × 2 m), and place them underwater within a lagoon in the Maldives. We then try to study the sand accumulation based on the changes in seasons and direction of the currents and waves. The most successful geometries, thus far, act like a shelf or reef, where the water and sand flow over the geometry, which creates turbulence and eddies,

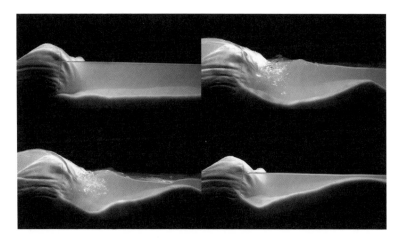

A series of lab experiments have been conducted in wave tanks studying the relationship between geometry, sand accumulation, and the wave characteristics. *Credit:* Self-Assembly Lab, MIT, Invena, Taylor Perron, James Bramante, Andrew Ashton, Tencate, SASe Construction, Planet, Vulcan, Allen Coral Atlas

promoting the sand to drop toward the sea floor. We are also gathering weather data and aiming to continually simulate the dynamics of the system computationally to inform new physical experiments in the lab and field tests. Our ultimate goal is to be able to understand the relationship between our successful tank experiments and the full-scale field experiments, and then deploy a number of successful geometries to the site in the Maldives.

This research will likely take us many years to achieve successful sandbars or island-scale accumulations, testing all of the variables and working toward a collective solution. We are now starting to see promising early results, with hundreds of cubic meters of sand accumulating in just a few months. We envision the long-term possibility of this project as a system that could help grow the islands in the Maldives, potentially overcoming the effects of sea-level rise, or rebuild beaches around the world that are rapidly eroding due to increasing storm inundation.

A series of lab experiments have been conducted in wave tanks studying the relationship between the sand formation patterns and the wave characteristics. *Credit:* Self-Assembly Lab, MIT, Invena, Taylor Perron, James Bramante, Andrew Ashton, Tencate, SASe Construction, Planet, Vulcan, Allen Coral Atlas

These submersible geometries could be deployed in different regions around the world and moved easily to adapt to the local seasons and wave forces. We are ultimately hoping this becomes a system that collaboratively works with the environment to provide a new approach to resilience and adaptation for coastal communities. Our multifaceted and collaborative approach with materials and the environment appears to be the only way to design for this extremely complex and ever-changing problem.

With this approach we aren't forcing materials into place and we aren't fighting the forces of the environment. We aren't predetermining the specifics of the design or the entire functionality of the system. We are working collaboratively and designing with materials and forces to evolve a solution that isn't predetermined. We are listening as much as we are telling the materials what to do. In this way surprising solutions can emerge, like the complex and beautiful patterns of sand ripples, or the native sandbar and island formations that happen naturally. Similarly, attempting to predict the shape and location of the ideal sustainable island or sandbar through dredging is extremely difficult. However, if we can promote the accumulation to happen on its own, then the hope is that the final output is far more sustainable and highly functional. In this way, the approach is a lot more like cultivating or collaborating than sculpting or constructing.

The vision, ultimately, is to arrive at a new kind of resilience—one in which destruction can also be construction. Think of the way that muscles build: it's tearing that prompts them to grow stronger. We now understand that forest fires—whether naturally occurring or artificially created—can promote a healthier and more diverse ecosystem over time. There are many instances where some form of destruction occurs that promotes the increasing strength, resilience, and construction of something else. Similarly, in coastal regions or islands, storm surges often bring with them massive amounts of sand, which end up accumulating on the beaches, sometimes making them bigger and healthier, while also promoting new growth of vegetation, which strengthens the foundations of the island. Our project demonstrates the potential to harness these forces in a productive way. Just as we discussed previously, things don't always need to fall apart—destruction can be construction, and it can be used as a way to promote resilience if we collaboratively tap into the natural agency of materials and their environment. We can't keep thinking that our top-down human designs will solve some of the most complex problems in our physical world; we need to work collaboratively with our environment.

Design from the Bottom Up

THROUGHOUT THIS BOOK I have discussed how materials can compute, communicate, make decisions, sense, and actuate. These active materials can be used for new forms of manufacturing, construction, robotics, products, and environments. Yet, as these capabilities are emerging and we are learning how to program materials, we need to understand how to design with and for materials. It is one thing to know that materials can be robots, or that we will one day have self-transforming products and environments. But how do we go about designing and creating these things?

The previous chapter explored how we can exploit the mechanisms of self-assembly and the properties of materials to build products, even environments, from the ground up. I argue that, in order to make the most of materials' potential, we must design from the bottom up as well, taking materials as our collaborators. Designing with active materials is not as easy as forcing a static material into place. Success requires a great deal of strategy and patience, and it involves changing the design process as well as the mindset of the designer—whether architect, product designer, or engineer. The traditional design process is top down and typically results in a product or structure that has only one main function or a limited life span. Ultimately, some finished products are made, and then recycled or (more often) simply thrown away. The process today can be summarized as:

Design > Build > Function > Dispose

The design process with active materials must change to create better products that won't contribute to our ever-increasing mountain of single-use and lifeless products. This process begins with the materials rather than the design itself. We will begin with the material building blocks and build from the bottom up. Designers will focus on the properties and structure of the material, augmenting them through new forms of fabrication in order to grow architectures with embedded material performance. This augmentation will mean that we'll infuse information, sensing, and actuation into the material parts and then allow design and functionality to emerge. This is similar to a traditional craft-based design process in which a woodworker or metalsmith has a deeply ingrained knowledge of the properties of a material, but our future design process takes this further, to enhance and augment materials to behave in unconventional ways, tapping into their natural ability while giving materials the agency to design along with us.

Design may still start with an application and a person—a "matter programmer," if you will—but then materials will adapt and continue to improve throughout their lifetime. The matter programmer may trigger the design process, but the material and its context of forces will design, evolve, become a product, and continue to transform into other products and functions based on need. In this new approach, design will be discovered and cultivated rather than simply realized. It will look something like this:

Material > Assembly > Design > Adaptation

This new approach amounts to a *collaboration* with the material, in which design and function emerge through the material's augmented behavior. It is a new development in the relationship between form and function, one where the form and function will change over time. This is a relationship where the form and function are symbiotically linked: as one adapts and changes, the other changes as well, influencing one another in a feedback loop. If we design specifically with materials, giving agency to their function and configurations, we can design for multiple functions and unforeseen future applications, and alleviate the necessity to have one final design. We will have multiple versions of the form and function.

The art world provides some early, useful examples of bottom-up design processes. Generative artists like Sol LeWitt, generative composers like Brian Eno, computational and information designers like Casey Reas and Ben Fry—all have used simple rules and relationships that were then propagated, iterated, and morphed to generate pieces of art from drawings, paintings, sculpture, or music. This type of conceptual and physical creation doesn't start from a vision for the final piece or from a brilliant idea; rather, the final piece emerges from the enactment of simple rules that collectively come together to create the complex final piece.

More recently, with the rapid advances of computational design tools, this bottom-up approach has exploded in architecture, product design, graphic design, and many other creative fields. The goal with these tools is to create a simple set of rules that generate a variety of possible design outcomes. In this way, generative tools act like a collaborative designer, working alongside human designers to propose solutions with many iterations and functional or formal variation. In a simplistic version, the goal of computational tools is to allow the design process to be much more about an exploration of what is possible. The "perfect" solution then may emerge on its own

Preprocess execution (also called deprocess). The text of a computer code (shown in gray) is connected by blue lines, drawn during the execution of the code. The generative image depicts a moment in time with the darker lines representing the more recent loops of the code. *Credit:* Ben Fry and Casey Reas, 2005, Archival pigment print on Hahnemühle photo rag, 11 × 36 in / 28 × 91.4 cm, https://benfry.com/deprocess/

through iteration and variations of all possible combinations. But sometimes there is no "perfect" solution, given that some design problems can be multivariable problems with many trade-offs for any given solution; thus, no single solution will be able to solve for all possible constraints. For example, the design of a component in a building, car, or plane, may have many constraints, from cost to weight, structural performance, or thermal performance, and each of those criteria are linked to one another. So the "perfect" solution in one category may make a worse solution in another, and thus, it may not be possible to make the "perfect" design decision across many competing constraints. In that case, a computationally bottom-up exploration can ensure that rather than a human designer trying to pick one solution, the collaboration can generate many solutions and evolve a design to maximize across many constraints. It could be time-consuming or even impossible to explore all possible designs if the designer has to do it manually, step by step, so instead designers over the past decades have been employing the collaborative capacity of computation to help with evolving bottom-up design solutions.

Recently, we are seeing the increasing application of machine learning with generative adversarial networks (GANs) to create new design possibilities.[1] By using existing datasets and competing neural networks to generate plausible yet completely made-up examples inspired by the original datasets. We have seen these tools used to generate realistic images of people's faces, architectural designs, animals, artwork, products, city maps, and even three-dimensional models. It is clear that these powerful tools will become increasingly prevalent in collaborations (or even completely autonomously) with human designers, generating alternative design scenarios that wouldn't have been conceived of previously. These may or may not solve for "perfectness," but they will certainly create strange, new and exciting alternatives to purely human imagination.

Architects throughout history like Antoni Gaudi or Frei Otto have explored bottom-up approaches in collaboration with materials, in the form of physical "form-finding" methods to arrive at informed design solutions. In their work, forces such as surface tension or gravity are made to act on a material, like a soap bubble or hanging chain, to create a novel, lightweight form that is often structurally efficient.[2] In this way, natural forces and their effect on materials become a design generator to show us novel form and function. Similarly, the lineage of Charles Darwin's *Origin of Species*, D'Arcy Thompson's *On Growth and Form*, and John Frazer's *An Evolutionary Architecture* leads to a new generation of possible form-finding techniques that don't require the slow evolution over many generations of species or many centuries of time passing.[3] We can now promote the evolution of form through dynamic forces in the environment to transform simple materials in real time. Since we can design and create new material structures, we can encode novel material behaviors in relation to forces in their environment. In this way, the future of design with active materials enhances the possibilities not only to "find" synthetic forms through natural forces but also to program behaviors and capabilities within materials that wouldn't naturally occur.

Contemporary form-finding techniques can be seen in the work of Achim Menges and his team at the University of Stuttgart's Institute for Computational Design (ICD), where they use wood veneer in precisely placed orientations to take advantage of the wood's natural anisotropic behavior and ability to curl when subject to moisture.[4] With wood grain patterns combined in different orientations, when the structure is left outside and humidity changes, the structure will form itself into various three-dimensional structures. In the most recent advances of

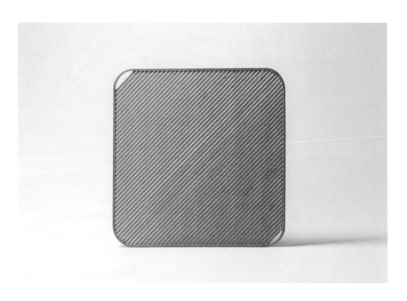

A process of 3D printing wooden structures with customized grain patterns promoting their physical transformation when subjected to moisture. The grain of the wood absorbs moisture, causing it to swell and create curvature in the wooden structure. Printing unique geometric patterns causes them to fold into precise angles or intricate designs like bowls and baskets. *Credit:* Self-Assembly Lab, MIT, and Christophe Guberan

this technique, our lab worked collaboratively with product designer Christophe Guberan as well as the ICD team to develop printed wooden structures where we could create customized wood grain patterns that wouldn't naturally form and promote unique transformations when subjected to moisture. For example, we created wood that can fold to 90-degrees on its own or a flat pattern that can morph into wooden bowls and baskets.[5]

In other examples, Andrew Kudless's or Mark West's work on fabric-formed concrete or plaster structures uses simple textiles that are stretched across a frame and then solidified into soft-looking forms after the cast material is poured.[6] When the plaster or concrete is poured into the textile it causes it to stretch and bulge due to the weight of the material. The final object emerges from the relationship among the plaster material, the effects of gravity, and the interaction of the stretched and constrained textile material. These beautiful examples are contemporary plays on form-finding processes that take advantage of simple materials and the forces of the environment to generate intricate, precise, and beautifully constructed structures that likely wouldn't have been generated with traditional machined fabrication processes or manual design tools.

We can now go beyond found form and create our own material compositions activated with different sources of energy—not just existing materials and singular forces. For example, a classic form-finding scenario is a simple chain that is hung from two points to create a catenary curve due to gravity. This curve is then flipped upside down to create an efficient structural arch. Flipping the model from hanging to standing, the forces acting on a simple material system inspire new and efficient designs. We can now go beyond this and create a chain that has programmed joints that encode a series of angles. For example, in a project at the Lab we printed a chain that

A series of components that can be connected in different patterns, then shaken randomly, and they will fold into precise three-dimensional structures based on the pattern of units. *Credit:* Skylar Tibbits, Neil Gershenfeld, Kenny Cheung, Max Lobovsky, Erik Demaine, Jonathan Bachrach, Jonathan Ward

had specifically designed joints such that when the chain was shaken, it could transform into the shape of a spiral, or a sinusoidal wave, or a geometric cube—depending on the code and the force of shaking.[7] We could easily change the pattern of joints, representing up, down, left, or right, and therefore get a completely different shape with the same random shaking. We could also change the design of the joints so that if there was a different force or frequency, the chain could take on a different shape. This programmed chain is made only of simple printed components, but by carefully designing the joints to encode different forms of logic we can embed new types of behaviors into something as simple as a hanging chain. By creating agency in materials, we can fabricate them with complex micro- and macrostructures that behave in ways that naturally found materials often cannot.

This leads us to a different approach to design. To visualize this, imagine a pile of bricks sitting in a plaza and a crowd of people who can pick up the pieces and put them together. Imagine there are no instructions and nobody telling them what to do. What will they build?

There are several different approaches they might take, which we can phrase in terms of existing types of design. Together they might build a random mess—a completely haphazard group of personality types, skill levels, and other factors will lead to *design by chaos*. They might accomplish a less messy outcome if they worked together and decided on a design to build together, *design by committee*. Design by committee is only marginally better than design by chaos because it tends to engage an averaging effect, a lowest common denominator of sorts, something everyone can agree on—which excludes marginal, surprising, or risky ideas. Great ideas and the worst ideas would likely be thrown out for being too extreme.

Another slightly different outcome would be if the single loudest or most dominant person leads the group and the rest follow to help build it—this is the very common *genius designer* scenario. We recognize *genius designers* (or strong-willed individual decision makers) who put forward their solutions for all sorts of applications. This scenario can be better than a committee decision because there is clarity, agency, direction, and strong intentions. They can take risks and offer surprising solutions that would likely be thrown out in the committee. But it is certainly not optimal. It will likely leave out overlooked but potentially "better" scenarios suggested by quieter or less dominant people.

As we saw, computational and generative design is powerful because it can generate many incredibly rich design options. Software can take into account multiple variables and constraints simultaneously. Powerful design, simulation, and optimization tools allow us to test and analyze millions of iterations and allow the design solution to *evolve* computationally and digitally to sophisticated levels. This scenario, *design as optimization*, is extremely powerful. It can encompass both the aesthetic and creative modalities in addition to functional characteristics, while also satisfying committees and managing the interests of the individual designers. This assumes, however, that a single design will be chosen and that the design process has an end state that leads to a functionally *optimal* solution. The main challenge with design as optimization is predicting and simulating *all* of the possible scenarios, forces, constraints, or applications that a particular design will face in the physical world. A building designed in the 1950s, for example, might be used years later for completely different purposes. Or an airplane component can be designed for certain conditions, but it would be difficult to imagine all possible flying scenarios or crash scenarios and their resultant forces and vectors. A piece of infrastructure may be designed for one purpose and then eventually changed with a different context, down the road. The initial design may have been optimal at the time, but it was limited by the knowledge of the time period or the imagined functional criteria. It reminds us that no static design can truly be optimal over time and over various changes in the environment.

There is an alternative to all of the methods of design that I've mentioned above—one that programmable and active materials uniquely make possible: *design as adaptation.* This method combines a computational approach with a material form-finding process to promote adaptation. Fabrication can now be seen as just the start of a product's life, not the end. Materials can come together and take advantage of external forces, internal changes, or environmental constraints to update their design on the fly—even after the product is already made—like the continual design and adaptation of the aerial self-organizing structures, or the sand formations in the Maldives that we discussed previously. Design and functionality can adapt indefinitely in response to demands over the life of the system. Materials can reconfigure themselves to offer new functionality or new aesthetic possibilities. This approach provides material agency while also allowing the design to continue to grow and change.

Let's think through the pile of bricks in the plaza again. The *design as adaptation* scenario implies that materials could *collaborate* with humans, make adjustments on the fly, and continually adapt based on external forces. The components might describe loading conditions as each brick is stacked on top of the next. As we saw in the previous chapter, this could give cues to the assembler as to where to place bricks to create a precise structure, or where to move some of them to make more optimal structural performance based on the bricks' feedback on structural forces. Other unique factors like the orientation of the bricks for solar exposure could lead to more informed structures, where the bricks could give instructions to the assembler, almost like a compass reading, and then guide the placement in a specific orientation. They could make local decisions and continual adjustments during the assembly process, helping to create more precision or better design decisions. Or perhaps the pile of bricks wouldn't help during the construction phase: the people would build whatever they like, and then the bricks could adjust and reconfigure themselves *after* the fact, in response to the building's structural/thermal/environmental performance over time. This scenario is a bottom-up solution where the materials and humans collaborate. This scenario is far better than *chaos*, better than a *genius designer* or what a *committee* could have imagined, and better than a fixed yet computationally *optimal* design—because it can grow, evolve, and change over time.

In the *design as adaptation* approach, an object is designed not at the beginning, but during the process of creation and throughout its life span. Take our bodies, for example. They do not have start and end points of design or fabrication. We might argue that birth is a form of design fabrication, but our bodies continue to grow, repair themselves, rebuild, adapt to the environment, degrade, and are eventually recycled back into fundamental matter. As the environment changes our bodies, it's difficult to distinguish design from adaptation. If you quickly ascend to the top of a mountain at high altitude you may get altitude sickness or feel shortness of breath. But if you live up there, or spend a longer period of time, your body will adapt and get acclimated to the environment. You could see this as a simple reaction, but it is also arguably a new design development and a powerful ability, where our bodies can adapt and change based on the surrounding environment. Most of today's material products can't do that, but they should—and if we learn to collaborate effectively with materials, they can.

Approaching design as adaptation will have far-reaching effects on the life of our products and how we relate to them. One thing that will change is when and how a material is set free in the real world. The moment when a product is deployed or released into the world or when a building welcomes its first occupants is usually a giant cliff with a leap of faith for any designer. This is the moment when the creation comes to be used and misused in the world. Having the ability to embed agency into materials and change the genetic makeup of our future material structures, however, enables us to rewrite this moment of translation continuously.

This is somewhat analogous to one of the latest trends in software development: to deploy a product early, get feedback, adjust its functionality, and update continually.

This approach has typically not been as well suited to hardware development, because hardware is much more difficult, expensive, and time-consuming to adjust after it's been built and launched. This is critical when it comes to applications like airplanes, infrastructure, or medical devices, where our lives are at risk if an implementation fails. The designers working within these domains typically use the following approach: they design, simulate, prototype, test, test again, test further, implement on small runs, analyze, and test again—until they are certain of success in the final product. This process dramatically increases the lead time for the design, development, and implementation of any new technology.

Material products that can adapt and adjust after shipping, while they are in use, could potentially speed up this process. Active materials could enable a faster release of a product by improving over time and adjusting for unforeseen circumstances. This could also mean less material or fewer components because rather than trying to design for every imaginable condition, we could have *less stuff* and do *more*, adapting to the fluctuating forces or future demands. Materials don't need to do *everything*—in fact, they may only be capable of doing one thing at a time. But we want them to be able to *transition* from one thing to another and to be extremely good at what they do at any given time. Most materials today don't even have the capacity to do more than one thing. On the other hand, some materials today can do a lot of things—yet many of them poorly.

Instead, we should be thinking more about adaptive systems. We want a product to be the best it can be. The multifunctional aspect doesn't mean we should take five products and slam them together. The process should be lean and agile, seamlessly adapting from one state to another. Or the function could even be the same—a bridge serves the same function, to get from point A to point

B, for example, but it could adjust in subtle ways, based on the amount of vehicles going over it, or based on the time of day and the changes in the weather, to continually optimize its structure in relationship to the demand.

As we explored in the previous chapter, the only way to design for these variable conditions is to relinquish a bit of control. The physical products we build today are made by adding together many materials and devices. It's difficult to think of a humanmade system that is more than the sum of its parts. If we remove components from a machine, it typically fails. If we add components, it acquires new functionality. Even planes, buildings, and space shuttles are really the sum of their parts. Yet bottom-up systems tend to be far more than the sum of their parts. The brain, for example, is far greater than what one individual neuron can do. If a brain is damaged or a significant part is removed, the brain can often adapt and continue functioning normally. These complex systems *adapt* their design over time: their material is their design and performance, and it continues to change indefinitely.

Ultimately, the design approaches that I've described in this chapter represent not just a practical change in the way we design, but a philosophical one in the way we build a collaborative relationship with materials. Rather than impose our will from the top down, we must work with materials as collaborators and cultivate their agency—recognizing that materials have their own intelligence, and we can tap into this to design better objects and environments.

We can see analogies to such an evolution in other fields. For example, if we look at the development of synthetic biology, we can go all the way back to natural evolution with mutations, to more intentional breeding for desired traits, to genetic engineering and then synthetic biology and the programmability of biological circuits, to more recent CRISPR technologies with gene editing, which is quickly evolving into machine learning for and with synthetic biology.[8] In this case, DNA is the programmable building block, allowing for the natural or synthetic evolution and reprogrammability of functional biological structures. Rather than being constrained to only naturally occurring biological functions, as was the case in previous decades with more primitive forms of breeding, we can now create novel behaviors by designing, editing, and working with the material functionality. This may sound scary to some, but it has been happening naturally through evolution since the beginning of life on earth. This development will become increasingly important and powerful, allowing for a new category of medicine with specifically tailored gene therapies to cure genetic diseases. In a similar way, programmable materials allow us not only to utilize existing material properties and their naturally occurring response to their environment, like form finding, but also to reprogram and create radically different material properties that will give materials unique behaviors that were previously impossible.

All of these historical paths of technological development are analogous to the development that we are seeing in material programming and point to the ever-changing role of the designer (just like the evolving roles of doctors, or the scientists developing new medicines, or the computer programmer and artificial intelligence). Designers, engineers, and makers have a long, complex, and evolving relationship with materials as their medium, progressing from accidental discoveries of material behavior, to the honing of material craft, all the way up to precise digital fabrication creating mass customized objects. This is now transitioning into information-rich materials, tunable and targeted to specific behaviors, creating ever more material agency. This relationship with materials has evolved from the cultivator and the curator of materials, into the prescriber of material information, and is now likely transitioning into the material collaborator or the orchestrator who creates guidelines and boundaries and seeds directions for growth.

Perhaps more like parenting than dictating, designing with materials is now about promoting the growth and development of the material in useful, healthy, information-rich ways. Material designers can't fully control the material's behavior at every moment, but they can set guidelines, establish boundaries, and hopefully work together to build something unique. Some of this is built in and some of it comes from external cues. The good news for material designers is that they have one advantage over parenting—they can easily change the material's genetic code; they can read it, change it, adapt it, and try it again without substantial risks. This is really our true role as designers when creating programmable materials—to create the initial code and to cultivate the external factors to promote its evolved behavior. Just as a child grows, changes, and adapts, both formally and functionally, our products in the future will evolve their form and function over time.

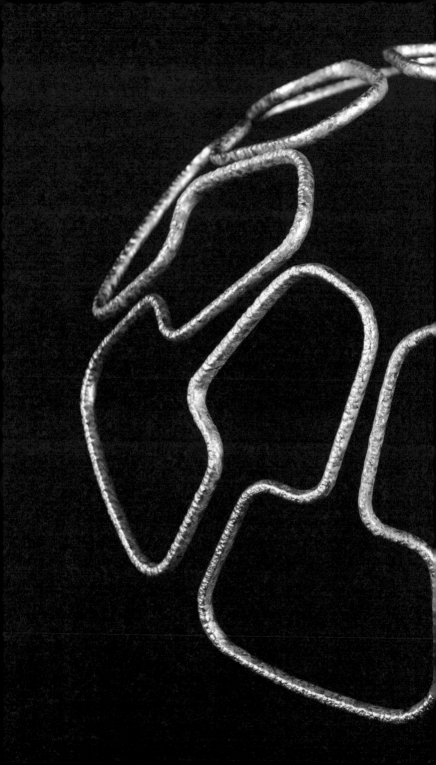

Reverse, Reuse, Recycle

ELECTRONIC DEVICES have taken over our environment, and they are contributing to the looming crisis of global waste. The increasingly short life spans of our electronic products lead to an endless cycle of waste—we end up purchasing new versions, with new materials, every few years, often discarding what is no longer up to date. We no longer think about our electronic devices getting simple hardware upgrades, with more memory, or a bigger battery, much less lasting longer, turning into better versions of themselves, or evolving into ever more usable products. Even if we take the time to install the latest software upgrade, our electronics often slow down or fall apart, and sometimes they are intentionally designed to slow down so that you'll buy the latest version.

It's not just consumer electronics. Children's toys, footwear, fashion, furniture, and many other industries are built on disposability, newness, and continually buying more. They typically operate with short life-span products and long life-span materials that pile up in landfills, seriously challenging the feasibility of recycling. It is time for us all to imagine more sustainable scenarios in which materials can be programmed so that products can be reversed, reused, changed, or adapted to new functions. When they stop working or aren't needed any longer, they can be broken down and transformed into something else.

Google, LG, and other companies took a step in this direction when they introduced the concept of modular phones in 2017.[1] The simple idea is that components could be added or removed from a "backbone" of the phone's core functions, which in turn would make the phone customizable and repairable. Consumers could build and customize the functionality of their phones by adding new features, multiple cameras, larger batteries, and so on. If a screen cracks, their modular phone components could be easily replaced by swapping out that component. In order to realize this idea, they also need to rethink the business models that surround the phone, which may be part of the reason that these concepts haven't taken off yet. The conventional notion is that a broken phone is good for business because consumers will simply buy a new one; however, companies need to grapple with the repercussions of additional waste.

Such visions of modularity and upgradable functionality are almost obvious as simple and effective strategies for increasing the life span of products. But visions such as these are not enough. It is one thing for tech companies to create a new device that can be easily repaired. But we need products that can repair themselves or at least be disassembled and reassembled into something completely new.

This vision of recyclability rests on several key concepts: customization, modularity, self-repair, disassembly, and growth. Customized products (instead of the typical mass production of standardized one-off goods) allow us to reduce waste directly by making only what's necessary, at the time it's needed. This would minimize the amount of unused inventory and reduce unnecessary production. We can also reduce the shipping, logistics, packaging, and energy needed to get manufactured goods to customers by creating localized and customized fabrication. We may not produce everything we need in our homes (although that is certainly one of the dreams that has been laid out for digital fabrication), but we could certainly produce more products regionally, on demand, and tailor production based on local need. Imagine manufacturing facilities in trailers traveling the country with reduced or limited raw material inventory and infrastructure, but capable of customized production. A number of companies are now starting to do this, like Patagonia's "Worn Wear" and mobile repair tour that traveled the country fixing people's damaged clothing. A mobile system of manufacturing and repair could also be oriented toward local creation and reusability, with products that can be designed, built up, and then disassembled and reconfigured into other functional objects.

Customization

Customization in product design is a contemporary trend due to the promise of digital fabrication, but it goes beyond just customizing the look and feel. As we have seen, we can customize material properties and functionality as well. One way to rethink a product's material performance is to simply reduce the amount of material it uses, which in turn can decrease its weight and increase its performance, in the case of automotive or aviation components. We can also, however, create active material products that do more with less. We can design structures that are smarter—that adapt, morph, and self-repair—so that products can grow stronger and more useful over time. This type of smart product would create a dramatic shift in our ways of thinking: it will change our mental reflexes about performance and disposable goods.

Advances in fabrication processes that customize not only the design of a product, but also its material and performance could also help reduce waste. Most of today's fabrication processes are blind, incapable of managing failures halfway through production, and they certainly can't tell if a part will be too fragile in its final implementation. Novel fabrication processes that are currently in development will be able to adapt, on the fly, by understanding successes, failures, and functional criteria. It could come to a halt and adapt when it detects a mistake. The process might adjust the material or fabrication parameters to make better parts and reduce part failures. For example, a printer could change the mixing ratios of a deposited material, on the fly, to adjust for stiffness, enhance transparency, or increase the bond of a material to its substrate. Wojciech Matusik's Computational Fabrication Group at MIT and their spin-out company Inkbit have developed a multimaterial printer that utilizes computer vision and machine learning to create a feedback loop, detecting mistakes and improving the printing process on the fly.[2] This process combines the fabricator and the quality-control expert into one superhuman (machine).

The ideal scenario would be a machine that could fully reverse an object, sucking it back up like a vacuum, and trying again (and again) until it emerges as imagined—or better than imagined. Taking one step toward this vision, in 2017 we developed a technology called rapid liquid printing (RLP) in collaboration with product designer Christophe Guberan and the furniture company Steelcase as a new method of printing that could promote rapid design, production, and perhaps, ultimately, reversibility.[3] The initial challenge was to address the three main limitations associated with 3D printers: their speed, their size, and the quality of the resulting products. Today's 3D printers are slow, small, and often have relatively weak material properties compared with traditional industrial manufacturing. We developed a process that prints within a gel suspension, creating large objects with a variety of industrial materials. The printed

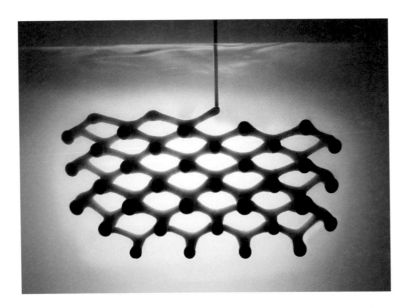

Rapid Liquid Printing: a new method of printing within a gel support environment enabling large-scale, fast production of high-quality rubbers, foams, and plastic objects. The gel provides support to the objects and enables fully three-dimensional tool paths for creating objects in any orientation. *Credit:* Self-Assembly Lab, MIT, Christophe Guberan, and Steelcase

material is extruded within the gel, which eliminates the effects of gravity on the object. It then cures with a chemical-curing process or UV light, and is removed from the gel as a solid object. By eliminating gravity, the process can print free-form in 3D space, rather than a series of stacked 2D layers that are created with a traditional 3D printing process. More like 3D calligraphy, RLP draws objects in 3D, and does not require extra printed material to support the 3D structure, as is normally the case with traditional 3D printing, because the gel suspends the object in space.

Similarly, with this process you do not need to wait for the material to cure at the moment of extrusion. Typically, in other forms of 3D printing, like fused deposition modeling (FDM) or stereolithography apparatus (SLA), the machine needs to fully cure or extrude and harden the material at every step along the printed path; otherwise, the part won't solidify and will sag. Obviously, this cure time dramatically reduces the speed of printing in traditional machines. Our system can print as fast as the machine can go, and the liquid material is cured after it is deposited. If there is an error, or the designer wants to make a change in the printed part, any uncured material can be suctioned back up, deposited somewhere else, and then allowed to cure. Once the prints are solidified, we simply take them out of the gel, wash them off with water, and they are ready to be used. A medical device, an orthotic, a fully formed shoe, a lamp (or various other printed parts that we have created) can be available in tens of minutes. It is an elegant, multimaterial printing process that works for products of many different sizes.

Our RLP process uses standard industrial materials such as foams, rigid plastics, rubbers, and we've even experimented with concretes. This range of industrial materials will take us one step closer to industrial production applications since traditional printing platforms have been limited by the quality of their materials. We recently printed inflatable structures to test the limits of our silicone rubber materials and high-resolution surfaces. The structures resemble balloons and soft pneumatic robots, yet we can print them with various complex geometries and internal chambers. We used a very flexible and stretchable silicone rubber, and we printed 3D structures that are airtight and watertight. These printed air pockets can theoretically be of any size or shape and have various internal or external geometries. Once the printed structure is removed from the gel, it can be inflated and the silicone can expand or create a morphable soft robot.

This printing process allows us to create robotic structures that are soft and inflatable. Traditionally, producing soft robotic structures requires manually casting molds with silicone rubber or creating complex designs from laminated sheets. The complexity and time-consuming nature of these fabrication techniques has limited the behaviors and capabilities of soft robotics to date. RLP makes it easier to produce much larger and more complex structures than would be feasible with a typical casting or lamination process. We have explored many applications, like adaptive cushioning that can morph to the body or a mobile robot walking across a table—all without any traditional electromechanical mechanisms. This fabrication technique could help accelerate the development of soft, multimaterial, and inflatable robots and will lead to many more ways to use active products.

Going beyond printing with silicone in the gel environment, more recently we developed a method to print metals.[4] Other researchers, like Michael Dickey at North Carolina State University, have been able to print low-temperature liquid metals with promising applications. However, our process focused on extruding high-temperature molten metals in three dimensions within a powder support bed to eliminate the effect of gravity.[5] This process has all of the same advantages of the gel in that we do not require printed supports or postprocessing, we can print very large structures, and it is even faster than the chemical curing process in RLP since the molten metal cools in a matter of seconds. This means that we can print metal objects from centimeters up to meters in a few minutes. The most compelling aspect of this printing capability, however, is the reversibility. We can quickly produce a metal component, and then, if it's not needed, we can directly melt it down again and print something else without any waste. This can happen over and over again, creating a waste-free metal production process.

Liquid metal printing: a new process for printing liquid metal within a powder support chamber enabling large-scale, fast production of high-quality metal objects. The nozzle moves in three-dimensional paths, extruding molten metal within the powder. This process is fully recyclable by melting down the parts and reprinting them within the powder chamber. *Credit:* Self-Assembly Lab, MIT, and AWTC

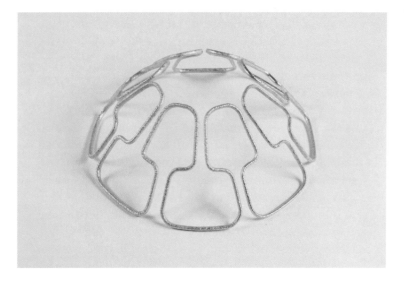

A metal bowl printed with liquid metal deposition within a powder support chamber. *Credit:* Self-Assembly Lab, MIT, and AWTC

These new fabrication processes are unlike traditional forms of manufacturing and greatly differ from more contemporary examples of additive manufacturing. With these approaches, we can create customized objects with high-quality material properties and new multimaterial functionality (like 4D printing or inflatable soft robots), while also reversing and undoing any unwanted designs. We can even make and unmake objects, creating a fully recyclable manufacturing pathway. In the near future customization and reversibility may be key aspects for redefining a sustainable path forward for design and manufacturing.

Modularity

Most products are made out of assemblies that include many materials and functional pieces. We can imagine a product made of components that can be built up into one function, and then broken down into constituent parts, and rebuilt into completely different functions. Take the cell phone that we discussed earlier, made of functional units—batteries, microprocessors, screens, buttons, microphones, and speakers, to name a few. These parts could be key components of many other electronic products. We could break down the products—not necessarily down to their chemical constituents, which may limit their reversibility, but to a macrolevel—so they could be accessed, stored, repurposed, and reassembled, allowing for infinite possibilities. This is the most basic form of modularity.

In the field of robotics, modular and self-reconfigurable robots have similar benefits in terms of recyclability.[6] With some number of functional building blocks, these robots can be built up, taken apart, and built into new functional devices, and they are infinitely reusable. This modularity can make design more distributed; with teams of people working, they can break down the tasks into specific functional building blocks. It also provides a platform for teams around the world, or kids to build their own custom robots. It ensures a longer life span for the robotic platform, rather than a one-off specialized robot that may be expensive to make, complicated to control, and only perform a specific task. As with Legos, there really isn't any waste. A family may buy a Lego kit with a specific theme, but after it is built and destroyed, all of those Legos can then be reused to build something else. That is one of the beautiful and sustainable aspects of Lego, that they can be anything and are used for years, even generations. (On the flip side, it is also one of the things people love to hate: that Legos seem never to go away and always to be sitting around waiting for someone to step on them!)

Many have highlighted the notion that there is no waste in a forest.[7] All living things are eventually broken down into their natural elements and recomposed into other living things. So if we could break down our manmade components into functional building blocks, we might be able to assemble and disassemble continuously to eliminate waste altogether. Forests and nearly all biological and chemical constructs are effectively modular and reconfigurable. Functional active materials may mean that in the future we can design this ability directly into our products.

Self-Repair

Instead of finding easier ways to swap out components or coming up with new ways that people can fix their own devices, we could aim to design parts that can repair themselves. Such products can incorporate material capabilities that can heal themselves—think of a scratch on your skin: the skin adapts to the injury and repairs the damaged region. There are a number of researchers exploring self-repairing composites, such as concretes, plastics, hydrogels, rubbers, and other materials that will soon find their way into everyday products. In concrete, researchers have developed a number of ways to produce self-healing properties, from microbes that generate calcite to polymers that fill in any structural cracks.[8] Chemically, much of the development in self-healing polymers comes from the ability of molecules to break and then rebond under certain conditions, essentially gluing themselves back together as if there had been no failure. Another group of researchers has developed a polymer material that can self-heal and even grow itself by consuming carbon dioxide from the air and energy from sunlight, like photosynthesis in plants.[9] All of these self-healing materials can form under some type of activation energy, like pressure, temperature, light, or others, and could be applied to things like self-healing bulletproof body armor, concrete for roads and buildings, and even body panels for planes or rockets.

As these materials develop further, perhaps we may no longer need to dispose of cracked screens in our cell phones, because our glass may be able to self-heal. Bridges could adjust their structure based on load and even heal the concrete and steel if either is damaged after an earthquake. Think of the millions of miles of road in the United States made of asphalt and concrete, which expand and contract due to dramatic changes in temperature and moisture, causing them to crack and require continual maintenance. In the near future, perhaps this infrastructure could heal on its own.

One common geometric motif that is used to create this kind of self-repair is to build in a bilayer structure: when the outer layer is punctured, the inner layer is activated and will expand or release new material to fill the gap. Think of a pipe that has two layers. When the outer layer is punctured, the inner layer is revealed, coming into contact with the air. This puncture could lead to a chemical reaction with the inner material, causing a foam or adhesive to expand and fill the hole. Rather than being severely compromised by an oil spill, a gas leak, or a massive flood, the infrastructure of the future may self-heal, creating safer and more resilient cities.

Disassembly

With self-assembly comes self-disassembly. If we know how a structure assembles, then we should know how it ought to disassemble under very specific conditions. We can create self-assembling devices that require vibration or rotation and then invoke self-disassembly by manipulating temperature, moisture, or pressure. In this way, we can design the input energy for disassembly to go far beyond the typical operational environment. By using temperatures that exceed humanly tolerable conditions, we could activate self-disassembly only at the highest or lowest extremes. In this way we can separate the functions of assembly and disassembly, enabling us to activate a set of materials to become a product or to break themselves down into simple building blocks on demand.

Research led by Dr. Joseph Chiodo looks at active disassembly for everyday products. For example, the team developed a simple bolt that can be used to adhere components together in a typical product assembly, but when heat is applied, the thread of the bolt disappears and the connection is broken.[10] The components easily slide off the bolt, and the assembly now comes apart into a number of individual pieces. This type of behavior can be used in many of today's products that rely on unique materials and components with complicated adhesion techniques from mechanical bonding to chemical bonding. All of the different materials make it extremely complicated to take things apart, recycle them, and reuse them at a later point. Self-disassembly can solve this problem.

Another approach to disassembly is disintegration, where materials can literally dissolve or otherwise fall apart on demand. As discussed earlier, Fiorenzo Omenetto's Silk Lab develops products that can be easily dissolved in a solvent, going from a physical optical device or a circuit, dissolving into a functionless liquid whenever needed. Similarly, John Rogers and other researchers have pioneered the field of "transient electronics" and devices that are "Born to Die."[11] Daniela Rus's research group at MIT has developed a small-scale robotic system that can self-assemble, transform then dissolve, and perform many other functions, yet it can quickly disintegrate in a solvent, leaving behind little to no waste. The broad goal of this dissolvable research is aimed at electronic devices, environmental sensors, or medical devices that can be implanted within the body and then dissolved when they are finished or are not

A small-scale material robot that walks, swims, and can even dissolve itself for recyclability. *Credit:* Photo by Christine Daniloff, MIT. Project by Shuhei Miyashita, Steven Guitron, Marvin Ludersdorfer, Cynthia Sung, Daniela Rus

needed any longer. Like stitches that dissolve themselves after a wound is healed, these devices can be designed to have sophisticated and adaptive functionality, and then programmed to dissolve.

At the construction scale, disintegration is also possible, as shown in our research on granular jamming with rocks and string that we discussed earlier. Related work by other researchers, such as Heinrich Jaeger from the University of Chicago or Achim Menges and Karola Dierich's aggregate construction techniques, has shown architectural-scale structures that can be rapidly built and unbuilt with simple materials through granular jamming.[12] The advantages of these simple material construction systems are plentiful from a construction standpoint—they are nearly instantaneously solid, with load-bearing capacity and no need for precise placement of components. The material can be poured like a liquid and you can use simple found materials. Yet these types of structures can also be easily disintegrated into a pile

The disassembly of the four-meter-tall tower built from only loose rocks and string using the principle of granular jamming. The tower was disintegrated by simply winding up the string to allow the tower to return to a pile of rocks and a spool of string. *Credit:* Gramazio Kohler Research, ETH Zurich, and Self-Assembly Lab, MIT

of raw materials and then rebuilt, over and over again. Almost like quicksand, the rock-based jammed structures can be disintegrated into loose rocks and a spool of string, then poured again into large load-bearing structures, and dissolved again.

Some might ask why we would want architectural structures that can be easily dissolved or disassembled. Doesn't that pose a risk? Of course, we don't want architecture or infrastructure to fall apart accidentally or structures to collapse easily. Yet, we also don't want to always build 100 percent permanent structures. Permanence is great if the structure can be used in many different circumstances, with many different functions, across decades or centuries. But structures that lose their function or value, or scenarios where we want to build something quickly that can also be quickly removed or changed, are great applications for disassembly. Think of disaster scenarios, or temporary structures for events, seasonal construction, or even structures that may be used for many years but then could be easily recycled into the next building. All of these examples are well suited to take advantage of reversible, self-disassembling material systems, whether that is dissolving materials or jammed structures that can transition from solid to liquid and back.

By observing the relationships among components, their geometry, and their physical interactions, we can understand what type of activation energy will bring them together and take them apart. This approach will change our perspective on recyclability from a purely chemical process, where we break down materials with limited cycles (think of plastics or paper, which typically can be functionally recycled only a finite number of times). We can create new ways to separate components and disassemble products. Imagine if all of our products were required to be reversible and their construction hinged on a two-way street: with a flip of a switch, manufacturers could instantly disassemble their products into the constituent parts. Imagine those materials reassembling into other products on their own, and self-disassembling as needed. That would have a substantial impact on the future of sustainability.

Growth

A final way that active materials may lead us toward a more sustainable future is through their ability to adapt and grow. For example, shoes that could increase comfort and support in the areas where you need it most. Or traction on a tire that gets better with wet conditions, and buildings or bridges that grow stronger with extreme loading conditions. These are all theoretically possible through material adaptation, though they will take time to fully develop.

Products that can adapt and grow in this way could potentially help create less waste, by reducing the amount of consumption, and reducing our impulse to quickly dispose of our goods. We could have materials that not only transform or assemble but literally grow and adapt to changing needs. Growth is a logical continuation in making smarter materials, an aspect that argues for not only reversibility but an alternative form of recyclability that allows things to grow and change over time. We often see this characteristic in old buildings that are adapted and reused for many different functions. In many ways, this shows an ability to grow with their tenants. We see this with humans and any living system that grows in physical size, mental capacity, and technical skill. Could this trait also be applied to the physical products and built environments around us? Could our materials help promote this type of renewal and evolving betterment as a replacement for the single-use and disposable culture that we are so entrenched in?

Recently, we have seen an influx of growing materials in product design and architecture. David Benjamin's installation at the Museum of Modern Art's PS1 Summer Pavilion in New York City was built with bricks grown from mycelium, the root structure of mushrooms.[13] Similarly, companies like Ecovative are literally growing a range of mushroom-based products from textilelike applications to building materials, lamps, and even packaging.[14] Mycelium-grown packaging has now been adopted in a wide range of industries aiming to replace the massive amount of waste from disposable packaging materials like Styrofoam. To produce mushroom-grown products, they create a mold in the shape of the product, then set up the right conditions for the mycelium, and then let it sit and watch it grow. This challenges our conventional processes of manufacturing, either additively placing materials or subtractively removing materials. It is more aligned with cultivating the environment and ingredients while promoting the material to grow into the shape and function of the product. In the case of mycelium, the material can even biodegrade or be composted. This is likely the most direct application of cradle-to-cradle manufacturing by promoting natural material growth.

Biofabricated materials could provide an alternative path to food, fashion, and products that consumers love without the harmful effects on the environment. Recently, we have seen the introduction of lab-grown meat and non-animal-based leathers challenging the meat and fashion industries.[15] Suzanne Lee's early work growing garments from cellulose is an example of what she calls "biocouture," evoking a future of "living factories," growing rather than manufacturing.[16] These technologies, although nascent, are tapping into the rapid advances in synthetic biology and materials science, utilizing material growth and translating those principles into functional objects. There is a long road ahead and

A denim jacket grown from cellulose rather than traditional garment manufacturing and cut-and-sew assembly methods. *Credit:* Biocouture Denim Jacket 2006, © Biofabricate 2020, https://www.biofabricate.co/

substantial challenges to compete with industrial manufacturing in terms of price, volume, and performance, but these examples are demonstrating early steps toward this next generation of biofabricated material goods. Through this work, growth can be seen as a novel technique for fabrication/manufacturing, but it can also be seen as an alternative to disposability. Many of these future material products could be grown, and then be composted or dissolved, reducing the waste that ends up in our landfills.

These products are mostly focusing on the beginning and end of the product life cycle—how we make and how we dissolve our physical products. By tapping into active materials and new ways of fabricating, however, we could promote materials to grow and change after they have been produced, as discussed in the previous chapter. Neri Oxman's Mediated Matter research group at MIT developed a series of silkworm pavilions, which were literally fabricated by worms.[17] They placed thousands of silkworms on a large three-dimensional scaffold, and then allowed the worms to extrude their silk, collectively weaving a silk coating around the space. These worms could be seen as a collaborative approach to construction, while also promoting the continual growth and adaptation of the environment. There are a number of other species that build things, from weaver birds to beavers, ants, and spiders. Human designers, too, could collaborate not only with the growth of materials, but also with the animal builders to constantly grow and adapt structures during their functional lifetime. These possibilities are likely to be more prevalent in the near future, with products being grown in a lab or a garden rather than a factory and then allowing for their continual transformation, growth, repair, and eventual decay when they are no longer needed.

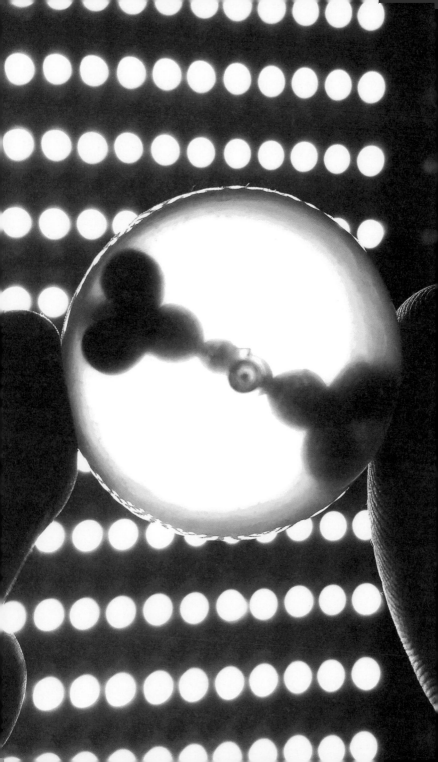

The Future of
Matter Is Evolving

IN THIS BOOK, I've outlined new ways to think about materials today, how we can interact with and program material behaviors to change our modes of creation, and how the performance of our products and environments will evolve in the future. I've described the opportunities that we create for ourselves when we collaborate with materials. Once we start to understand novel ways for materials to behave and to function, we can start to see how design, fabrication, and environmental life cycles change.

I have also illustrated the software and hardware revolutions that have led to this materials revolution by shifting individual makers and industrialized producers toward material hackers and matter programmers. The intersections advancing knowledge from materials science, computer science, artificial intelligence, and synthetic biology are rapidly converging. This is leading to unprecedented new material characteristics capable of producing matter that is at once programmable *and* highly active.

I believe that virtually everything we design and create as humans and the natural environments that we occupy could potentially change in the coming decades due to the new ideas and radical advances in materials I've described in this book. The products we use will become active and adaptive, assembling themselves, responding to changes in the surrounding environment, and evolving to meet our needs. We can remake the product life cycle, designing and collaborating with materials in unique ways that shift our focus from static products and environments to highly active, lifelike behaviors. Our manufacturing tools and processes will shift with emerging platforms for digital fabrication and new capabilities for materials to grow and self-assemble on their own. Shipping and logistics will likely change with more distributed capabilities and packages that pack flat, and then transform, expand, and adapt on the other side. We will transition from a system of disposability and simplistic models of recyclability, to one that centers around customization, modularity, self-repair, disassembly, and growth.

Traditional distinctions in our physical environment between manmade and natural—between buildings and infrastructure, and landscapes and geography—may erode, blending into a symbiotic collaboration between the natural and the synthetic. These two don't need to be at odds; they can work and evolve together. Our landscapes will continue to change and adapt, much like the weather, the seasons, and the tides, but we can learn to work with these changes and guide them toward adaptation for positive change, rather than wait for destruction or try to control nature. We will soon realize that building our manmade cities and static infrastructure will need to shift into a model that is more resilient and adaptive than our current seawalls, dredging, barriers, and bridges.

Our perspectives on products are literally being reprogrammed. We should embrace the affordances that these new technologies make possible through material agency. Today we have the opportunity to rewrite our relationship with the physical world so that we are no longer passive observers but active collaborators. Understanding the design principles for programming materials will help catalyze materials into doing more than ever before: to think on their own, and to act, grow, respond, and make decisions. This future should bring true material agency.

But we are still in the early stages of creating this program-mable landscape. Drawing an analogy with computer science, we are in the days of 0s and 1s and mainframe terminals. We've shown that materials can compute, ac-tuate, and sense, but we haven't yet built the sophisticated interfaces and the elegant embodiments, or implemented the seamless applications of the future. We need to trans-late these experiments, lab-scale demonstrations, early prototypes, and experimental product applications into something more ubiquitous, scalable, widely understood, and applicable. It isn't solely a technological scale-up problem, however. This is also about technological adop-tion curves, and we are currently sitting in the innova-tors and early adopters phase. In order to advance, there needs to be a careful collaboration among the material designers as inventors, the industrial implementers, and the everyday users as adopters. The first market applica-tions will help cross the chasm from early adopters to the majority. These initial applications will help gain wider understanding and attention and likely trickle down to other sectors.

We need to level up to enable the applications that can make our everyday lives better. To do so, researchers, policymakers, and the public alike will need to continue to change our perception about what is possible and what is necessary, imagining and realizing new possibilities. Instead of seeing problems and thinking of impossibili-ties or, alternatively, imagining top-down machines and energy-intensive robots as the solution, we should be thinking about materials and how to push the boundaries between design and science—seeing the possibilities that active materials can reveal.

Blending of Disciplines

We've seen in this book how programmable materials bring together design with highly technical and typically specialized fields like computer science, materials science, biology, chemistry, and engineering. This hybridization is enabling entirely new design processes, material advances, and fabrication capabilities like multimaterial printing, whole-garment knitting, and rapid liquid printing, all of which were either previously impossible or relegated to specialized trades and outside of a designer's purview. The design process is changing rapidly to encompass novel methodologies that have been ingrained in other fields, like network behaviors, complex systems, evolutionary capabilities, and artificially intelligent architectures. Designers of active matter ultimately need to be "material whisperers," imagining with their minds and hands, while taking those capabilities beyond material craft into material agency. Training the matter programmers of the future will thus require that STEM become STEAM, as many have suggested before—that we add the arts to science, technology, engineering, and math.[1]

The beginnings of this change are already afoot. Here at MIT, for example, I've observed increasing numbers of mechanical engineering, computer science, materials science, biology, and other science or engineering students enrolled in design courses. Conversely, we are also seeing an increasing number of design students taking science and engineering courses. These students recognize that the future lies at the intersection of science, engineering, and design. The demand has been so substantial that we've created entirely new design minor and major degrees that enable students to blend the creative with the technological, and to graduate as polymath designers of the future—computer science designers, biologist designers, physicist designers, planetary designers, environmental designers, or any other type of hybrid designer. Elsewhere, hybrid masters of design programs at Carnegie Mellon, Berkeley, and various other universities similarly highlight this hybridization of disciplines as the next evolution of design.[2] This shift in design is exactly the embodiment and ethos that led to programmable materials, the blending of the computational and physical design with the engineering of materials.

We'll need to create the right environment to foster this way of thinking and working. Throughout this book, I've described how programming matter demands a mix of radical and creative exploration with relevant and technical developments. This approach is predicated on rigorous play, experimentation, and honing one's craft, learning from materials while scientifically testing, developing, and inventing. In the same way that the revolution in personal computing was led by people creating personalized and creative applications for desktop computers—not just efficient bookkeeping or mathematical models, as originally predicted—the realization of smarter materials systems will likely come from the creative customization and playful realization enabled by today's fabrication. It requires an environment that operates and feeds on opposing forces—freedom and constraint, creativity and intellect, abstraction and reality, research and application. Progress is driven by a generative balance between possible applications that inspire and give us direction, and enough creative space to think big, beyond incremental advances and band-aid fixes.

This type of environment has precedents: think of Bell Labs throughout the twentieth century, IBM, Microsoft, Apple, and Xerox in the 1970s, 1980s, and 1990s, or Google X today. But the speed at which we now move is making it increasingly difficult to find and cultivate these types of environments. The success of software technologies coming out of Silicon Valley and the speed of rapidly changing information have focused our attention on the present, on immediacy, on "nowness," as well as on fastest-to-market and fastest-to-product tunnel vision—as opposed to deep and fundamental change. Timelines on deliverables are reinforcing more conservative and incremental development instead of encouraging creative and radical long-term solutions. There are other factors at play as well: financial uncertainty and government shutdowns, pandemics, climate change, and many other looming scenarios, which have challenged traditional funding for research. These shortages shift research toward short-term applications, which in turn often leads to more problem solving instead of longer-term advances and radical thinking. Given the societal shortsightedness, it is now even more important that we find the freedom to blend creative far-future possibilities with near-term applications and engineered developments through active materials.

Common Platform

In order to further advance this newly emerging hybrid discipline, we need to create a common language and strategy that designers and scientists share. Around the world, there are scientists, engineers, and designers creating active materials with various design tools, working on many different materials (wood, metals, plastics, rubbers, foams, and so on) with many different fabrication processes (printing, knitting, lamination, injecting, and so on), each with unique activation mechanisms and testing strategies. On the one hand, this explosion of research and technological capabilities is important and exciting for the field. On the other hand, this diversity also creates challenges in communication, repeatability, knowledge transfer, and acceleration when researchers do not share a common language or toolset. We need to establish the fundamentals and science behind programmable materials: a language, a platform, protocols, all of which will work together toward a common strategy for activating matter. We need to create the software and hardware applications that constitute a common language for programming materials, used across disciplines, institutions, and regions. We need standards in the process, metrics, and reliable testing procedures for how we design, build, and analyze active materials. And even more important to help accelerate the development, we need a more unified strategy to share information across research groups, disciplinary boundaries, and fabrication platforms.

An ideal experimental platform would allow designers to fabricate a material structure with various properties and activation energies. Digital inputs would allow for various activation energies using electricity, temperature, moisture, light, pressure, and others. We could analyze and record the behavior using image/video tracking, force measurements, and other physical metrics. The process of recording and analyzing will help identify new capabilities, gaps in the research, successful approaches, and failed attempts. And we could share our conclusions with the rest of the world, seeking feedback and guidance for potential applications and risks. Much like an open-source platform, citizen science project, or many other shared systems, this common platform would advance the community, technology, and ethics of its development. Such a platform for experimentation would help shape a common programming language for materials, and a kind of periodic table for active matter.

This creation of a common language and approach could then both increase the speed of discovery and facilitate necessary conversations with everyday users and policymakers about potential applications as well as risks. As with many technological advances, the shift to active matter may happen right before our eyes and still remain barely noticeable. Ubiquitous computing and the internet of things are good examples of this phenomenon. We generally still live the same lives, and yet computing devices are ever-present in our homes, offices, cars, gyms, streets, and outer space. Artificial intelligence, too, is already embedded in our daily lives, even though the humanoid robot with general intelligence is nowhere to be seen. It is estimated that there are thousands of processes that we interact with every day that are powered by narrow artificial intelligence that go almost unnoticed, like streaming products and music recommendations, air traffic control systems, autonomous driving and flight systems, Google maps, email responses, and many others. Active matter will likely be similarly embedded in our everyday lives in seamless and elegant ways, often without even knowing it.

These parallels make clear the importance of a society-wide dialogue that keeps pace with technological developments. Active matter will create new risks and responsibilities that we must collectively confront as these capabilities take shape. We are moving toward a future where material products can literally evolve on their own, and we should certainly expect vigorous debates about ethical considerations that arise whenever new technologies and production methods challenge the status quo. For all of the rewards of active matter, what might be the potential risks? What is our role and responsibility in the life span of evolving and intelligent products? What is the role of the designer and what are the implications for manufacturing, jobs, and safety when materials change on their own? Who is responsible if something goes wrong—the designer, the fabricator, or the material itself? How do we create safety standards around highly active, ever-changing material products?

A common platform and language could help create a system of "checks and balances" by orchestrating close relationships among those who create new technologies, the public that uses the technologies, and those who govern. It would facilitate the sharing of designers' and scientists' conclusions with the rest of the world, seeking feedback and guidance from everyday users and policymakers. A continuing discussion among these three parties would allow the conversation and debate to adjust as technology evolves. There are disastrous examples of times when one of these parties has dominated the others or a new technology was debated behind closed doors. The Manhattan Project highlights the problems with closed-door debates, as the proper regulations and global implications of nuclear weapons were hardly examined publicly. Blockchain, artificial intelligence, autonomous vehicles, and gene editing similarly illustrate the diverse and divisive ways in which we adopt new technologies, and how they influence policy after the fact, each with positive and negative examples of technology, policy, and public impact. We need to encourage our societies to embrace and not fear technological progress, yet we need to foster invention with responsible applications and compassionate policies. A common platform can ultimately help us share information and standard practices not only to advance the capabilities but also to help guide ethical and responsible applications in the future.

If we can successfully change our perspectives toward what is possible and what is best to do with this new material agency, the future will be incredibly exciting. The creation of ubiquitous active materials will enable the personalization and democratization of smarter products with unique capabilities for manufacturing. The entire product life cycle will change with materials that can be assembled and disassembled, reconfigured, and adapted into various products and functionalities. These might be material products that are now enhanced with unique properties and performances or entirely novel product categories. We can imagine existing products that behave in surprising ways, or are manufactured with bottom-up pathways. Or we can imagine entirely new categories of materially intelligent products and environments, adapting, changing, and evolving. Many of these we can't foresee until we actually meet them in the future.

As artificial intelligence, industrial automation, and robotics become more pervasive, we will continue to question our human and robot relationships. Throughout this book, I've shown that the robots of the future may be very different from the robots of today. They may be soft, agile, squishy, flexible, adaptable, material robots. Tomorrow's robots will not only consist of materials that can sense, actuate, and compute, but they will be lighter, have fewer components, need less battery power, and cost less. These soft-material robots may make traditional electromechanical robots obsolete, or the soft-material robotics may become promising collaborators.

Although I have argued throughout this book that materials may replace our classical notion of electronic computing and mechanical robots, a very realistic near-term scenario is the tightly coupled collaboration with traditional forms of electronic computing, robotics, and active materials. This three-way collaboration among designers, materials, and robots will come with responsibilities, which point to interesting overlaps and differences in the ways in which we harness a material's agency. Humans can bring creativity, play, and collaboration to find better design solutions. Traditional electromechanical robots, for their part, are better equipped at global communication, fast computational power, and tireless precision or repetition. Materials can sense and respond to their physical surroundings in ways that humans and robots often cannot. This layered collaboration can lead to solutions that are creative and novel, precise and repeatable, and environmentally transformative.

Self-Replication

Once our materials become true collaborators, they will likely go beyond just interacting with humans and robot partners. Materials may also generate and pass information to new offspring. In self-replication, information is translated from one generation to another. Conventionally, this phenomenon is thought of mostly in three domains: the biological, the computational, and the robotic. Yet the physical and material domain of self-replication is emerging.

In the biological sense, we might think of replication as species that can replicate to produce offspring chemically and biologically. Replication is fundamental to life. Researchers have gone beyond biological replication to show computational replication as well, in which code can self-replicate and produce "offspring" code, filled with similarities and mutations just like its biological counterpart. Think of a software virus that can self-replicate, mutate, and propagate new strands of itself. Scientists have also focused more recently on robotic replication, in which a robot is capable of picking parts of itself from an inventory and building an exact copy.[3]

Long before the advances in computational and robotic replication, however, the British geneticist and mathematician Lionel Penrose created a series of videos called *Mechanics of Self-Reproduction* in 1958.[4] In this work, he produced ingeniously designed and crafted wooden blocks that could transmit information and propagate new offspring structures when they were agitated. The blocks were placed on a track, which was shaken back and forth. As the blocks bumped into one another, the shape and pattern of the initial blocks would promote connections, orienting neighboring blocks to either accept or reject future connections. These simple interactions could generate patterns and produce copies of an initial input. His work was a successful construction of a nonbiological self-reproducing object, and it has continued to influence and inspire us more than a half century later.

A system developed at our lab further illustrates the concept of self-replication through nonbiological and nonrobotic components. We used simple, hollow, plastic spheres to create the experiment, and we placed geometric structures of metal spheres and magnets inside.[5] The shape of the metal-and-magnet structure inside the plastic spheres allowed for flexibility and connectivity because the plastic spheres could rotate and connect to neighboring units, while the flexible metal-and-magnet balls inside them allowed the angle of the connection to change slightly. When the angle became too extreme, the connection broke, and the neighboring spheres separated. We made hundreds of these spheres and placed them on a vibrating table to give them the energy to move around, find one another, connect, and disconnect. Our goal was to observe the patterns that would emerge through the connected spheres and demonstrate cell-like structures that could grow and divide.

Initially, the spheres self-assembled into linear strands. Then, the strands eventually closed into circular structures. These circular structures grew to the point where the entire circle started to wobble so much that the angle of the units forced the spheres to break apart, in turn breaking the circle into two smaller circles. The circular structures that were the strongest were made of 5, 6, 7, or 8 spherical units, whereas the less stable structures with more than 10 spheres eventually separated. The

A series of physical spheres oscillating on a table demonstrates the process of self-replication including growth, encapsulation, and division. This project demonstrates cellular division through simple material components rather than robotic or biological systems. *Credit:* Self-Assembly Lab, MIT

circle structures exhibited growth and division when they grew to a certain size, and then split and divided into two smaller, more stable circles. Some circles also encapsulated other spheres when they formed around a single unit. This project showed us that simple nonbiological, nonelectronic, or nonrobotic mechanisms can visually behave like biological cellular growth and division.

Material replication without biological or robotic means is of particular interest in manufacturing. If we can create physical components that can self-reproduce continually in manufacturing, then we can create a machine that is capable of producing more machines—and we can eliminate the scalability challenges that plague traditional factories. Biology, as an example, is far more efficient and capable than our human systems of manufacturing. The sheer complexity of a human body, built from DNA all the way up, is said to have the complexity of Avogadro's number, which we have yet to build in any human construct.[6] Yet, the error rate in human DNA is roughly 10^{-8}, while our buildings, cars, planes, and other humanmade products continue to be plagued with high failure rates and low yields.[7] But with self-replicating materials, we could create material building blocks that produce other building blocks that potentially have built-in error correction and scalability.

Recent government programs like the Molecular Foundry and Atoms to Products (A2P), for instance, both focus on manufacturing from the bottom up, from molecules all the way up to products. They focus on chemical and biological structures that can self-assemble, grow, and achieve macroscale functionality. The aim is to promote radical advances in manufacturing capabilities through self-assembly and self-replication.[8] The dream is that these types of self-replicating facilities can be set up and then run themselves.

Material replication may seem like a dream in today's world of manufacturing. Perhaps it would be too slow, or by losing control in the design process it would perhaps lead to undesirable outcomes. But this is where materials can play a role in design, helping to create a natural evolution of higher performing, more unusual or novel functional characteristics in manufacturing.

Mass Speciation

The next step beyond material replication would be to transmit information to each generation, much like speciation. Once you have continuous mass production with replication, you can imagine creating differentiation, mutation, and evolution with each generation. The parts would likely have a fitness criterion and influence from the physical environment—to be stronger, lighter, more flexible, or higher-performing—which could produce offspring parts with encodings of their own fitness, so that over time, they could literally evolve their design or functionality.

From a manufacturing perspective, mass speciation would follow from mass production and mass customization capabilities. But this new manufacturing frontier would be about more than creating slightly different versions with the same properties and different looks; it would be about the evolution of intelligent material products into different products, functions, or systems over time. Just as we discussed earlier, there are analogies with polymorphism in biology, chemistry, or computer science, where the same building blocks or genetic code can mutate into completely different entities and functions without changing the underlying code. Think of graphene, graphite, and diamond, which are all made of carbon but have entirely different properties. Our new material building blocks could be fabricated with basic building blocks of code and functionality, yet will evolve into entirely different species depending on their context and environment.

Throughout this book, I've shifted our traditional perspective on the future of manufacturing to one that doesn't just rely on better robotic automation or cheaper labor. In a future driven by a collaboration with active materials, manufacturing will take advantage of this intelligence. Manufacturing will be more like breeding different species of material products. We will likely manufacture products that change over time to acquire different functionalities and continue to improve. We could be breeding different products with the same DNA but different personalities, performance, functions, and appearance emerge. The production process could be the same, but the entire life of the product could adapt and evolve. These systems may get better over time or, equally important, they may disassemble and become other things, evolving through dissolving.

Material AI

Once we have created fabrication systems with behaviors of mass speciation, these elements will likely inspire successive generations with higher levels of functionality. By embedding information directly into materials, we may be able to kickstart their literal evolution, creating new functional species, passing information from one generation to the next. As I showed with the example of aerial components that developed the functional criteria of flight, we should eventually go far beyond this example into true evolutionary adaptation. Each generation can respond to both internal and external fitness criteria, generating new behaviors or functionality. This type of material evolution can create new design possibilities and allow materials or products to improve over time. Evolutionary materials have, for the most part, been realized today only within biological or computational realms but could soon lead to major changes within the design and production of synthetic materials. We could be spawning lifelike and bottom-up species of materials that radically change the way we manufacture and grow our goods. This revolution will come from the materials—not the machines. We will embrace a new relationship with our products, more like a parent, a chef, or a gardener's goal of cultivating, rather than forcing, growth over generations.

If we look at the trajectory of computing, we see analog mechanical computers leading to digital logic and digital computing; then to personalized computing with agile and sophisticated advanced software and networks; and eventually to specialized intelligence, autonomous systems, and broader general intelligence capabilities. Maybe the same will be true of materials. Research into designing programmable materials is relatively new: first we had analog materials; that led to the first steps in developing digital materials and new material capabilities as digital fabrication was popularized. This paved the way for today's programmable materials with sophisticated capabilities to sense, actuate, compute, reconfigure, error correct, and interact in complex ways. We could very well be moving toward intelligent and autonomous materials that have lifelike properties and can execute more than simple codes or sophisticated programs. The next generation may well be a material AI.

As we have seen, sometimes materials can work better than traditional forms of computing or robots. Slime mold maze solvers, for example, have shown us that some of the most basic organisms can demonstrate powerful computational capabilities and decision making through spatial search strategies. They can easily navigate a complex environment to find food or draw out optimal networks for highway and subway infrastructure, often achieving these feats with more efficiency than our human approaches.[9] Scientists have also demonstrated that plants can predict the direction of sunlight, going beyond pure phototropism to associative learning much like animals.[10] It might appear from these examples that only living species can learn or demonstrate intelligence. Computational AI, however, demonstrates nonliving, nonbiological learning with specialized or maybe even general intelligence.

Why shouldn't inanimate materials be able to learn and have enhanced intelligence? Humans and all other biological systems are made out of simple materials, without any robots or computers, and we have grown to evolve human-level intelligence. Could other simple material combinations do the same? Can a piece of wood, metal, or stone predict patterns, learn, and adapt? Can we develop a material structure that can be designed, fabricated, and programmed to exhibit intelligence? This future of material intelligence may tell us more about our own human intelligence and how we can further augment our synthetic and digital forms of intelligence.

Creating more innately intelligent physical materials increases the odds that we can solve even greater challenges collaboratively. We will need to work together with materials to solve conceptual, mathematical, and computational challenges in the future. Just as we have used animals (pigeons for long-distance communication, dogs to sniff out drugs, rats to sniff out landmines, dolphins to detect underwater devices, pigs to find truffles in a forest) or even simple materials (mercury to measure temperature, magnets to indicate cardinal directions, and bimetals to regulate temperature in an engine), we should work with more intelligent materials in the future for design, human ingenuity, and problem solving.[11] Recall that in the 1700s, John Harrison brilliantly realized that he could use material properties as an elegant solution to a major global problem that people had thought would only be solved by complex astronomy, mathematics, or even magic. Similarly, the future of material intelligence may blend materials science, computation, and new fabrication capabilities to offer solutions to some of our most challenging problems today. As we are already seeing, the development of synthetic biology and gene editing technologies is promising to tackle some of the greatest threats, like cancer and genetic diseases.[12] Soon, everyday materials could think on their own, sense diseases, uncover hidden forces in our environment, act with autonomy, agency, and creativity, and help us solve a generation of global challenges.

This new medium of matter is active, fluid, and ever-changing. It has more in common with the fluctuations and power of the weather, the growth and adaptation of plants, the creativity and abstraction of humans, and the replication of any living species than it does with our traditional relationship to materials and robots of the past. Programmable materials may go beyond computing to become ubiquitously embedded in every molecule, crystal, strand, fiber, sheet, and block of material that we use to create the physical world around us. The new perspectives I've outlined in this book are critical to the ways matter programmers of the future will design and manufacture in this evolving materials revolution. These design principles reach beyond traditional notions of creating beautiful products or highly functional devices. They allow for design and manufacturing processes to converge, coalesce materially, and create lifelike products that adapt and evolve. To get there, it will be essential to rethink the design process: manufacturing and functionality will need to converge with ideation and creation, not remain separated like today's linear process of idea, then fabrication, then functionality. Fabrication, design, and material adaptation need to be one and the same. As the material structure is emerging, design and functionality are evolving. Manufacturing will be about designing, cultivating, and growing. Design will not be about ideas alone; it will be about collaborating with materials, understanding their potential capabilities, and hearing what functionalities or behaviors they have to offer. The material and the design process will be symbiotic.

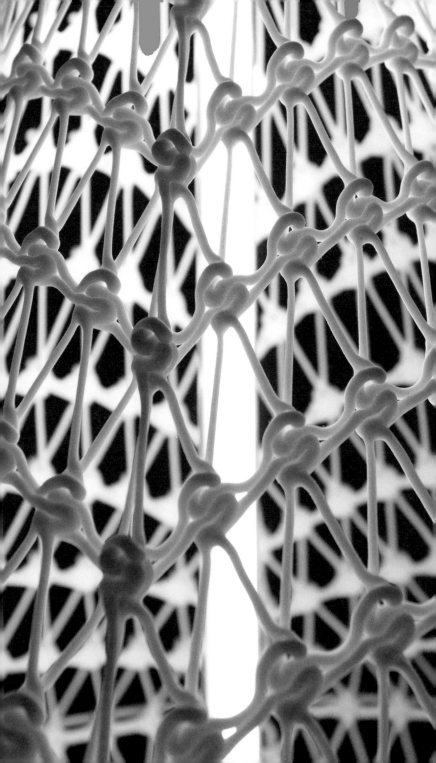

Notes

Programming Matter

1. ("Longitude: The True Story of a Lone Genius Who Solved the Greatest Scientific Problem of His Time" 1996)
2. I founded the Self-Assembly Lab in 2013, and it has since expanded to include directors Jared Laucks and Schendy Kernizan, and a diverse group of undergraduates, graduate students, and full-time researchers from many disciplines. The lab is part of the Department of Architecture and is housed in MIT's International Design Center.
3. In 2015, we organized the Active Matter Summit at MIT, bringing together artists, designers, scientists, and engineers working from a diverse range of fields. Subsequently, MIT Press published *Active Matter*, which includes a variety of work highlighting the advances of active matter across length scales. (Tibbits 2017)
4. (Goldstein, Campbell, and Mowry 2005)
5. (Stephenson 1998; Hall and Williams 2014; Cameron and Hurd 1984)
6. (Hans 1938)
7. (Howe 1955)
8. (Sutherland 1964)
9. ("Active Textile Tailoring" n.d.; "Climate-Active Textiles" n.d.)

Computing Is Physical

1. ("Arduino—Home" n.d.; "Makey Makey" n.d.; "STEM Kits & Robotics for Kids | Inspire STEM Education with Sphero" n.d.)
2. (Shetterly 2018)
3. (Shapiro 1987)
4. (Shannon 1940)
5. (McCartney 1999; Hodges 2014)

6. (Brinkman, Haggan, and Troutman 1997)
7. (Gershenfeld 2005)
8. (Popescu 2007)
9. (Negroponte 1996)
10. (Paul 2006)
11. (Davis 2018)
12. (Paun, Rozenberg, and Salomaa 2005; Church, Gao, and Kosuri 2012)
13. (Boneh et al. 1996)
14. (Church and Regis 2014)
15. (Chowdhury et al. 2019)
16. (Moorman 2020)
17. (Menges 2012; "Radical Atoms | ACM Interactions" n.d.)
18. (Prakash and Gershenfeld 2007; Katsikis, Cybulski, and Prakash 2015)
19. (Stanford University 2015)
20. (Fredkin and Toffoli 1982)
21. (Tibbits 2010)
22. (Karban 2015)
23. (Giraldo et al. 2014)
24. (Boley et al. 2019)
25. (Denning et al. 1989)
26. (Montfort 2016; Maeda and Burns 2005; Resnick and Robinson 2017; "Professor Seymour Papert" n.d.; Reas and Fry 2007)
27. (Reas and Fry 2007; https://processing.org/)
28. (Cardelli and Wegner 1985; Ford 1966)
29. (LeWitt et al. 2000; Steadman 2012)
30. (Turing 1952)
31. (Wolfram 2002)

Order from Chaos

1. (Schlangen and Joseph 2009; Blaiszik et al. 2010)
2. (Wei, Dai, and Yin 2012; Ke et al. 2012)
3. (Tibbits 2012)
4. (Tibbits and Falvello 2013; Olson 2015)
5. (Knight, Jaeger, and Nagel 1993)
6. (Papadopoulou, Laucks, and Tibbits 2017a)
7. (Vella and Mahadevan 2005)
8. ("Polymagnets—Correlated Magnetics" n.d.)
9. (Pray 2008)

Less Is Smart

1. (Senatore et al. 2017)
2. (Gambhir 2013)
3. (Aejmelaeus-Lindström et al. 2017; Aejmelaeus-Lindström et al. 2016)
4. (Cohen et al. 2020)
5. (Papadopoulou, Laucks, and Tibbits 2017b)
6. (Self-Assembly Lab et al. n.d.)
7. (Yao et al. 2015)
8. (Kim et al. 2010)

Robots without Robots

1. (Cheung et al. 2011; Knaian et al. 2012)
2. (Zhang, Demir, and Gu 2019)
3. (Morrison et al. 2015)
4. (Self-Assembly Lab et al. n.d.)
5. (Cramer et al. 2019; Jenett et al. 2017)
6. (Onal et al. 2015)
7. (Miyashita et al. 2015)
8. (Wehner et al. 2016)
9. (Tessler et al. 2020)
10. (Sanan, Lynn, and Griffith 2014)
11. (Harper 2014)

Build from the Bottom Up

1. (Marsh and Onof 2008)
2. (Melenbrink et al. 2017)
3. (Rubenstein, Cornejo, and Nagpal 2014)
4. (Willmann, Gramazio, and Kohler 2015; Gramazio, Kohler, and Langenberg 2017; Mirjan et al. 2016)
5. (Lussi et al. 2018)
6. (Lafreniere et al. 2016)
7. ("Beyond Vision" n.d.)
8. (Papadopoulou, Laucks, and Tibbits 2017a)
9. (Papadopoulou, Laucks, and Tibbits 2017a)
10. (Pilkey and Cooper 2012)

Design from the Bottom Up

1. (Goodfellow et al. 2014; Radford, Metz, and Chintala 2015; Karras et al. 2017; Vasu, Madam, and Rajagopalan 2018; Wu et al. 2016)
2. (Liddell 2015; Meissner and Möller 2015; Huerta 2006)
3. (Darwin 1859; Thompson 1917; Frazer 1995)
4. (Correa et al. 2015; Grönquist et al. 2019; Krieg et al. 2017; Menges and Reichert 2015; Reichert, Menges, and Correa 2015)
5. (Correa et al. 2015)
6. (Kudless et al. 2011)
7. (Tibbits et al. n.d.)
8. (Barrangou 2015)

Reverse, Reuse, Recycle

1. (Laser 2016)
2. (Sitthi-Amorn et al. 2015)
3. (Hajash et al. 2017)
4. (Self-Assembly Lab n.d.)
5. (Ladd et al. 2013)
6. (Moubarak and Ben-Tzvi 2012; Stoy, Brandt, and Christensen 2010)
7. (Gershenfeld, Gershenfeld, and Cutcher-Gershenfeld 2017)
8. (Blaiszik et al. 2010; Jonkers 2007; Zavada et al. 2015; Wu, Cai, and Weitz 2017; McMillan 2017)
9. (Kwak et al. 2018)
10. (Chiodo and Jones 2012)
11. (Chang et al. 2017)
12. (Murphy et al. 2016; Dierichs, Schwinn, and Menges 2013; Dierichs and Menges 2016)
13. ("Living Matter" 2017)
14. ("Ecovative Design" n.d.)
15. ("Modern Meadow" n.d.)
16. (Lee 2005; "Biofabricate" n.d.)
17. (Oxman et al. 2014)

The Future of Matter Is Evolving

1. (Land 2013)
2. ("Carnegie Mellon School of Design" n.d.; "Berkeley Arts + Design Programs" n.d.)

3. (Zykov et al. 2005)
4. (Penrose 1958)
5. (Papadopoulou, Laucks, and Tibbits 2017a)
6. (Papadopoulou, Laucks, and Tibbits 2017a; Gershenfeld 2005)
7. (Pray 2008)
8. ("Program Seeks Ability to Assemble Atom-Sized Pieces into Practical Products" 2015; "Molecular Foundry" n.d.)
9. (Reid et al. 2012; Evans 2020)
10. (Gagliano et al. 2016)
11. (Braverman 2012; Mehlhorn and Rehkämper 2009; Sasaki and Biro 2017)
12. (Cai et al. 2016)

References

"Active Textile Tailoring." n.d. Accessed June 15, 2020. https://self assemblylab.mit.edu/active-textile-tailoring/.

Aejmelaeus-Lindström, Petrus, Ammar Mirjan, Fabio Gramazio, Matthias Kohler, Schendy Kernizan, Björn Sparrman, Jared Laucks, and Skylar Tibbits. 2017. "Granular Jamming of Loadbearing and Reversible Structures: Rock Print and Rock Wall." *Architectural Design* 87 (4): 82–87.

Aejmelaeus-Lindström, Petrus, Jan Willmann, Skylar Tibbits, Fabio Gramazio, and Matthias Kohler. 2016. "Jammed Architectural Structures: Towards Large-Scale Reversible Construction." *Granular Matter* 18 (2): 28.

"Arduino—Home." n.d. Accessed June 16, 2020. https://www.arduino.cc.

Barrangou, Rodolphe. 2015. "The Roles of CRISPR-Cas Systems in Adaptive Immunity and Beyond." *Current Opinion in Immunology* 32 (February): 36–41.

"Berkeley Arts + Design Programs." n.d. Accessed June 15, 2020. https://artsdesign.berkeley.edu/design/programs.

"Beyond Vision." n.d. Marcelo Coelho. Accessed June 15, 2020. https://cmarcelo.com/work/beyond-vision.

"Biofabricate." n.d. Accessed June 15, 2020. https://www.biofabricate.co.

Blaiszik, B. J., S. L. B. Kramer, S. C. Olugebefola, J. S. Moore, N. R. Sottos, and S. R. White. 2010. "Self-Healing Polymers and Composites." *Annual Review of Materials Research* 40 (1): 179–211.

Boley, J. William, Wim M. van Rees, Charles Lissandrello, Mark N. Horenstein, Ryan L. Truby, Arda Kotikian, Jennifer A. Lewis, and L. Mahadevan. 2019. "Shape-Shifting Structured Lattices via Multimaterial 4D Printing." *Proceedings of the National Academy of Sciences of the United States of America* 116 (42): 20856–62.

Boneh, Dan, Christopher Dunworth, Richard J. Lipton, and Jiří Sgall. 1996. "On the Computational Power of DNA." *Discrete Applied Mathematics* 71 (1): 79–94.

Braverman, Irus. 2012. "Passing the Sniff Test: Police Dogs as Surveillance Technology." August. *Buffalo Law Review*. https://papers.ssrn.com/abstract=2142530.

Brinkman, William F., Douglas E. Haggan, and William W. Troutman. 1997. "A History of the Invention of the Transistor and Where It Will Lead Us." *IEEE Journal of Solid-State Circuits* 32 (12): 1858–65.

Cai, Liquan, Alfred L. Fisher, Haochu Huang, and Zijian Xie. 2016. "CRISPR-Mediated Genome Editing and Human Diseases." *Genes and Diseases*. https://doi.org/10.1016/j.gendis.2016.07.003.

Cameron, James, and G. Hurd. 1984. *The Terminator* [Motion Picture]. Orion Pictures.

Cardelli, Luca, and Peter Wegner. 1985. "On Understanding Types, Data Abstraction, and Polymorphism." *ACM Comput. Surv.* 17 (4): 471–523.

"Carnegie Mellon School of Design." n.d. Accessed June 15, 2020. https://www.design.cmu.edu/.

Chang, Jan-Kai, Hui Fang, Christopher A. Bower, Enming Song, Xinge Yu, and John A. Rogers. 2017. "Materials and Processing Approaches for Foundry-Compatible Transient Electronics." *Proceedings of the National Academy of Sciences of the United States of America* 114 (28): E5522–29.

Cheung, K. C., E. D. Demaine, J. R. Bachrach, and S. Griffith. 2011. "Programmable Assembly with Universally Foldable Strings (Moteins)." *IEEE Transactions on Robotics* 27 (4): 718–29.

Chiodo, Joseph, and Nick Jones. 2012. "Smart Materials Use in Active Disassembly." *Assembly Automation*. https://doi.org/10.1108/01445151211198683.

Chowdhury, Sreyan, Samual Castro, Courtney Coker, Taylor E. Hinchliffe, Nicholas Arpaia, and Tal Danino. 2019. "Programmable Bacteria Induce Durable Tumor Regression and Systemic Antitumor Immunity." *Nature Medicine* 25: 1057–63.

Church, George M., Yuan Gao, and Sriram Kosuri. 2012. "Next-Generation Digital Information Storage in DNA." *Science* 337 (6102): 1628.

Church, George M., and Ed Regis. 2014. *Regenesis: How Synthetic Biology Will Reinvent Nature and Ourselves*. Basic Books.

"Climate-Active Textiles." n.d. Accessed June 15, 2020. https://selfassemblylab.mit.edu/climateactive-textiles.

Cohen, Zach, Nathaniel Elberfeld, Andrew Moorman, Jared Laucks, Schendy Kernizan, Douglas Holmes, and Skylar Tibbits. 2020. "Superjammed: Tunable and Morphable Spanning Structures through Granular Jamming." *Technology | Architecture + Design* 4 (2): 211–20.

Correa, David, Athina Papadopoulou, Christophe Guberan, Nynika Jhaveri, Steffen Reichert, Achim Menges, and Skylar Tibbits. 2015. "3D-Printed Wood: Programming Hygroscopic Material Transformations." *3D Printing and Additive Manufacturing*. https://doi.org/10.1089/3dp.2015.0022.

Cramer, Nicholas B., Daniel W. Cellucci, Olivia B. Formoso, Christine E. Gregg, Benjamin E. Jenett, Joseph H. Kim, Martynas Lendraitis, et al. 2019. "Elastic Shape Morphing of Ultralight Structures by Programmable Assembly." *Smart Materials and Structures* 28 (5): 055006.

Darwin, Charles. 1859. *On the Origin of Species by Means of Natural Selection, or The Preservation of Favoured Races in the Struggle for Life.* W. Clowes and Sons.

Davis, Martin. 2018. *The Universal Computer: The Road from Leibniz to Turing.* 3rd edition. CRC Press.

Denning, P. J., D. E. Comer, D. Gries, M. C. Mulder, A. Tucker, A. J. Turner, and P. R. Young. 1989. "Computing as a Discipline." *Computer* 22 (2): 63–70.

Dierichs, Karola, and Achim Menges. 2016. "Towards an Aggregate Architecture: Designed Granular Systems as Programmable Matter in Architecture." *Granular Matter*. https://doi.org/10.1007/s10035-016-0631-3.

Dierichs, Karola, Tobias Schwinn, and Achim Menges. 2013. "Robotic Pouring of Aggregate Structures." *Rob | Arch 2012*. https://doi.org/10.1007/978-3-7091-1465-0_23.

"Ecovative Design." n.d. Accessed June 15, 2020. https://ecovativedesign.com.

Evans, Claire. 2020. "Beyond Smart Rocks: It's Time to Reimagine What a Computer Could Be." *Grow*. https://www.growbyginkgo.com/2020/07/15/beyond-smart-rocks/.

Ford, E. B. 1966. "Genetic Polymorphism." *Proceedings of the Royal Society of London. Series B, Containing Papers of a Biological Character. Royal Society* 164 (995): 350–61.

Frazer, John. 1995. *An Evolutionary Architecture.* Architectural Association Publications.

Fredkin, Edward, and Tommaso Toffoli. 1982. "Conservative Logic." *International Journal of Theoretical Physics* 21 (3): 219–53.

Gagliano, Monica, Vladyslav V. Vyazovskiy, Alexander A. Borbély, Mavra Grimonprez, and Martial Depczynski. 2016. "Learning by Association in Plants." *Scientific Reports*. https://doi.org/10.1038/srep38427.

Gambhir, Murari Lal. 2013. *Concrete Technology: Theory and Practice*. Tata McGraw-Hill Education.

Gershenfeld, N. 2005. *Fab: The Coming Revolution on Your Desktop*. Basic Books.

Gershenfeld, Neil, Alan Gershenfeld, and Joel Cutcher-Gershenfeld. 2017. *Designing Reality: How to Survive and Thrive in the Third Digital Revolution*. Basic Books.

Giraldo, Juan Pablo, Markita P. Landry, Sean M. Faltermeier, Thomas P. McNicholas, Nicole M. Iverson, Ardemis A. Boghossian, Nigel F. Reuel, et al. 2014. "Erratum: Corrigendum: Plant Nanobionics Approach to Augment Photosynthesis and Biochemical Sensing." *Nature Materials*. https://doi.org/10.1038/nmat3947.

Goldstein, S. C., J. D. Campbell, and T. C. Mowry. 2005. "Programmable Matter." *Computer* 38 (6): 99–101.

Goodfellow, Ian J., Jean Pouget-Abadie, Mehdi Mirza, Bing Xu, David Warde-Farley, Sherjil Ozair, Aaron Courville, and Yoshua Bengio. 2014. "Generative Adversarial Networks." http://arxiv.org/abs/1406.2661.

Gramazio, Fabio, Matthias Kohler, and Silke Langenberg. 2017. "Foreword by the Editors." *Fabricate 2014*. https://doi.org/10.2307/j.ctt1tp3c5w.3.

Grönquist, Philippe, Dylan Wood, Mohammad M. Hassani, Falk K. Wittel, Achim Menges, and Markus Rüggeberg. 2019. "Analysis of Hygroscopic Self-Shaping Wood at Large Scale for Curved Mass Timber Structures." *Science Advances* 5 (9): eaax1311.

Hajash, Kathleen, Bjorn Sparrman, Christophe Guberan, Jared Laucks, and Skylar Tibbits. 2017. "Large-Scale Rapid Liquid Printing." *3D Printing and Additive Manufacturing* 4 (3): 123–32.

Hall, D., and C. Williams. 2014. *Big Hero 6* [Motion Picture]. Walt Disney Studios.

Hans, Stoehr. 1938. Jacquard loom. USPTO 2136328. U.S. Patent, filed November 12, 1937, and issued November 8, 1938. https://patentimages.storage.googleapis.com/55/f3/4c/17fe047df6cc18/US2136328.pdf.

Harper, Robert. 2014. "Structure and Efficiency of Computer Programs." *Structure*. http://www.cs.cmu.edu/~rwh/papers/secp/secp.pdf.

Hodges, Andrew. 2014. *Alan Turing: The Enigma: The Book That Inspired the Film The Imitation Game*. Updated edition. Princeton University Press.

Howe, H. E. 1955. "Teaching Power Tools to Run Themselves." *Popular Science*. August.

Huerta, Santiago. 2006. "Structural Design in the Work of Gaudí." *Architectural Science Review*. https://doi.org/10.3763/asre.2006.4943.

Jenett, Benjamin, Sam Calisch, Daniel Cellucci, Nick Cramer, Neil Gershenfeld, Sean Swei, and Kenneth C. Cheung. 2017. "Digital Morphing Wing: Active Wing Shaping Concept Using Composite Lattice-Based Cellular Structures." *Soft Robotics* 4 (1): 33–48.

Jonkers, Henk M. 2007. "Self Healing Concrete: A Biological Approach." *Springer Series in Materials Science*. https://doi.org/10.1007/978-1-4020-6250-6_9.

Karban, Richard. 2015. *Plant Sensing and Communication*. University of Chicago Press.

Karras, Tero, Timo Aila, Samuli Laine, and Jaakko Lehtinen. 2017. "Progressive Growing of GANs for Improved Quality, Stability, and Variation." http://arxiv.org/abs/1710.10196.

Katsikis, Georgios, James S. Cybulski, and Manu Prakash. 2015. "Synchronous Universal Droplet Logic and Control." *Nature Physics* 11 (7): 588–96.

Ke, Yonggang, Luvena L. Ong, William M. Shih, and Peng Yin. 2012. "Three-Dimensional Structures Self-Assembled from DNA Bricks." *Science* 338 (6111): 1177–83.

Kim, Dae-Hyeong, Jonathan Viventi, Jason J. Amsden, Jianliang Xiao, Leif Vigeland, Yun-Soung Kim, Justin A. Blanco, Bruce Panilaitis, Eric S. Frechette, Diego Contreras, David L. Kaplan, Fiorenzo G. Omenetto, Yonggang Huang, Keh-Chih Hwang, Mitchell R. Zakin, Brian Litt, and John A. Rogers. 2010. "Dissolvable Films of Silk Fibroin for Ultrathin Conformal Bio-Integrated Electronics." *Nature Materials* 9 (6): 511–17.

Knaian, A. N., K. C. Cheung, M. B. Lobovsky, A. J. Oines, P. Schmidt-Neilsen, and N. A. Gershenfeld. 2012. "The Milli-Motein: A Self-Folding Chain of Programmable Matter with a One Centimeter Module Pitch." In *2012 IEEE/RSJ International Conference on Intelligent Robots and Systems*, 1447–53. Institute of Electrical and Electronics Engineers.

Knight, J. B., H. M. Jaeger, and S. R. Nagel. 1993. "Vibration-Induced Size Separation in Granular Media: The Convection Connection." *Physical Review Letters* 70 (24): 3728–31.

Krieg, Oliver David, Zachary Christian, David Correa, Achim Menges, Steffen Reichert, Katja Rinderspacher, and Tobias Schwinn. 2017. "HygroSkin." *Fabricate 2014*. https://doi.org/10.2307/j.ctt1tp3c5w.37.

Kudless, Andrew, Urs Leonhard Hirschberg, Martin Kaftan, and Roberto Apéstigue García. 2011. "Bodies in Formation: The Material Evolution of Flexible Formworks." https://pdfs.semanticscholar.org/8732/b25b7b8ceddcb4d609a71f51401102fae245.pdf.

Kwak, Seon-Yeong, Juan Pablo Giraldo, Tedrick Thomas Salim Lew, Min Hao Wong, Pingwei Liu, Yun Jung Yang, Volodymyr B. Koman, Melissa K. McGee, Bradley D. Olsen, and Michael S. Strano. 2018. "Polymethacrylamide and Carbon Composites That Grow, Strengthen, and Self-Repair Using Ambient Carbon Dioxide Fixation." *Advanced Materials* 30 (46): e1804037.

Ladd, Collin, Ju-Hee So, John Muth, and Michael D. Dickey. 2013. "3D Printing of Free Standing Liquid Metal Microstructures." *Advanced Materials* 25 (36): 5081–85.

Lafreniere, Benjamin, Tovi Grossman, Fraser Anderson, Justin Matejka, Heather Kerrick, Danil Nagy, Lauren Vasey, et al. 2016. "Crowdsourced Fabrication." *Proceedings of the 29th Annual Symposium on User Interface Software and Technology*. https://doi.org/10.1145/2984511.2984553.

Land, Michelle H. 2013. "Full STEAM Ahead: The Benefits of Integrating the Arts into STEM." *Procedia Computer Science*. https://doi.org/10.1016/j.procs.2013.09.317.

Laser, Stefan. 2016. "A Phone Worth Keeping for the Next 6 Billion? Exploring the Creation of a Modular Smartphone Made by Google." *Müll*. https://doi.org/10.14361/9783839433270-009.

Lee, Suzanne. 2005. *Fashioning the Future: Tomorrow's Wardrobe*. Thames and Hudson.

LeWitt, Sol, Martin Friedman, San Francisco Museum of Modern Art, and Whitney Museum of American Art. 2000. *Sol LeWitt: A Retrospective*. Yale University Press.

Liddell, Ian. 2015. "Frei Otto and the Development of Gridshells." *Case Studies in Structural Engineering*. https://doi.org/10.1016/j.csse.2015.08.001.

"Living Matter." 2017. *Active Matter*. https://doi.org/10.7551/mitpress/11236.003.0040.

"Longitude: The True Story of a Lone Genius Who Solved the Greatest Scientific Problem of His Time." 1996. *Choice Reviews Online*. https://doi.org/10.5860/choice.33-3880.

Lussi, Manuel, Timothy Sandy, Kathrin Dorfler, Norman Hack, Fabio Gramazio, Matthias Kohler, and Jonas Buchli. 2018. "Accurate and Adaptive in Situ Fabrication of an Undulated Wall Using an on-Board Visual Sensing System." *2018 IEEE International Conference on Robotics and Automation (ICRA)*. https://doi.org/10.1109/icra.2018.8460480.

Maeda, John, and Red Burns. 2005. "Creative Code." *Education* 7: 177.

"Makey Makey." n.d. Makey Shop. Accessed June 15, 2020. https://makey makey.com/.

Marsh, Leslie, and Christian Onof. 2008. "Stigmergic Epistemology, Stigmergic Cognition." *Cognitive Systems Research*. https://doi.org /10.1016/j.cogsys.2007.06.009.

McCartney, Scott. 1999. *ENIAC: The Triumphs and Tragedies of the World's First Computer*. Walker & Company.

McMillan, Fiona. 2017. "The Rise of Self-Healing Materials," December. https://www.forbes.com/sites/fionamcmillan/2017/12/21/the -rise-of-self-healing-materials/.

Mehlhorn, Julia, and Gerd Rehkämper. 2009. "Neurobiology of the Homing Pigeon—a Review." *Naturwissenschaften*. https://doi.org /10.1007/s00114-009-0560-7.

Meissner, Irene, and Eberhard Möller. 2015. *Frei Otto: A Life of Research, Construction and Inspiration*. Birkhauser.

Melenbrink, Nathan, Panagiotis Michalatos, Paul Kassabian, and Justin Werfel. 2017. "Using Local Force Measurements to Guide Construction by Distributed Climbing Robots." *2017 IEEE/RSJ International Conference on Intelligent Robots and Systems (IROS)*. https://doi.org/10.1109/iros.2017.8206298.

Menges, Achim. 2012. *Material Computation: Higher Integration in Morphogenetic Design*. John Wiley & Sons.

Menges, Achim, and Steffen Reichert. 2015. "Performative Wood: Physically Programming the Responsive Architecture of the Hygro-Scope and HygroSkin Projects." *Architectural Design*. https://doi.org /10.1002/ad.1956.

Mirjan, Ammar, Federico Augugliaro, Raffaello D'Andrea, Fabio Gramazio, and Matthias Kohler. 2016. "Building a Bridge with Flying Robots." *Robotic Fabrication in Architecture, Art and Design 2016*. https://doi.org/10.1007/978-3-319-26378-6_3.

Miyashita, S., S. Guitron, M. Ludersdorfer, C. R. Sung, and D. Rus. 2015. "An Untethered Miniature Origami Robot That Self-Folds, Walks, Swims, and Degrades." In *2015 IEEE International Conference on Robotics and Automation (ICRA)*, 1490–96. Institute of Electrical and Electronics Engineers.

"Modern Meadow." n.d. Accessed June 15, 2020. http://www.modern meadow.com/.

"Molecular Foundry." n.d. Accessed June 15, 2020. http://foundry.lbl.gov/.

Montfort, Nick. 2016. *Exploratory Programming for the Arts and Humanities*. MIT Press.

Moorman, Andrew. 2020. "Machine Learning Inspired Synthetic Biology: Neuromorphic Computing in Mammalian Cells." Master's thesis, Massachusetts Institute of Technology.

Morrison, Robert J., Scott J. Hollister, Matthew F. Niedner, Maryam Ghadimi Mahani, Albert H. Park, Deepak K. Mehta, Richard G. Ohye, and Glenn E. Green. 2015. "Mitigation of Tracheobronchomalacia with 3D-Printed Personalized Medical Devices in Pediatric Patients." *Science Translational Medicine* 7 (285): 285ra64.

Moubarak, Paul, and Pinhas Ben-Tzvi. 2012. "Modular and Reconfigurable Mobile Robotics." *Robotics and Autonomous Systems*. https://doi.org/10.1016/j.robot.2012.09.002.

Murphy, Kieran A., Nikolaj Reiser, Darius Choksy, Clare E. Singer, and Heinrich M. Jaeger. 2016. "Freestanding Loadbearing Structures with Z-Shaped Particles." *Granular Matter*. https://doi.org/10.1007/s10035-015-0600-2.

Negroponte, Nicholas. 1996. *Being Digital*. Vintage Books.

Olson, Arthur J. 2015. "Self-Assembly Gets Physical." *Nature Nanotechnology* 10 (8): 728.

Onal, C. D., M. T. Tolley, R. J. Wood, and D. Rus. 2015. "Origami-Inspired Printed Robots." *IEEE/ASME Transactions on Mechatronics* 20 (5): 2214–21.

Oxman, Neri, J. Laucks, M. Kayser, J. Duro-Royo, and C. Gonzales-Uribe. 2014. "Silk Pavilion: A Case Study in Fibre-based Digital Fabrication." In *FABRICATE Conference Proceedings*, ed. Fabio Gramazio, Matthias Kohler, and Silke Langenberg, 248–55. gta Verlag.

Papadopoulou, Athina, Jared Laucks, and Skylar Tibbits. 2017a. "From Self-Assembly to Evolutionary Structures." *Architectural Design* 87 (4): 28–37.

———. 2017b. "General Principles for Programming Material." In *Active Matter*, 125–42. MIT Press.

Paul, Chandana. 2006. "Morphological Computation: A Basis for the Analysis of Morphology and Control Requirements." *Robotics and Autonomous Systems* 54 (8): 619–30.

Paun, Gheorghe, Grzegorz Rozenberg, and Arto Salomaa. 2005. *DNA Computing: New Computing Paradigms*. Springer Science & Business Media.

Penrose, L. S. 1958. "Mechanics of Self-Reproduction." *Annals of Human Genetics* 23 (1): 59–72.

Pilkey, Orrin H., and J. Andrew G. Cooper. 2012. "'Alternative' Shoreline Erosion Control Devices: A Review." In *Pitfalls of Shoreline Stabilization: Selected Case Studies*, edited by J. Andrew G. Cooper and Orrin H. Pilkey, 187–214. Springer.

"Polymagnets—Correlated Magnetics." n.d. Correlated Magnetics. Accessed June 15, 2020. http://www.polymagnet.com/polymagnets/.

Popescu, George A. 2007. "Digital Materials for Digital Fabrication." Master's thesis, Massachusetts Institute of Technology. http://hdl.handle.net/1721.1/41754.

Prakash, Manu, and Neil Gershenfeld. 2007. "Microfluidic Bubble Logic." *Science* 315 (5813): 832–35.

Pray, Leslie. 2008. "DNA Replication and Causes of Mutation." *Nature Education* 1 (1): 214.

"Professor Seymour Papert." N.d. Accessed June 15, 2020. http://www.papert.org.

"Program Seeks Ability to Assemble Atom-Sized Pieces into Practical Products." 2015. Accessed June 15, 2020. https://www.darpa.mil/news-events/2015-12-29.

Radford, Alec, Luke Metz, and Soumith Chintala. 2015. "Unsupervised Representation Learning with Deep Convolutional Generative Adversarial Networks." http://arxiv.org/abs/1511.06434.

"Radical Atoms | ACM Interactions." n.d. Accessed June 15, 2020. https://interactions.acm.org/archive/view/january-february-2012/radical-atoms.

Reas, Casey, and Ben Fry. 2007. *Processing: A Programming Handbook for Visual Designers and Artists*. MIT Press.

Reichert, Steffen, Achim Menges, and David Correa. 2015. "Meteorosensitive Architecture: Biomimetic Building Skins Based on Materially Embedded and Hygroscopically Enabled Responsiveness." *Computer-Aided Design*. https://doi.org/10.1016/j.cad.2014.02.010.

Reid, Chris R., Tanya Latty, Audrey Dussutour, and Madeleine Beekman. 2012. "Slime Mold Uses an Externalized Spatial 'Memory' to Navigate in Complex Environments." *Proceedings of the National Academy of Sciences of the United States of America* 109 (43): 17490–94.

Resnick, Mitchel, and Ken Robinson. 2017. *Lifelong Kindergarten: Cultivating Creativity through Projects, Passion, Peers, and Play*. MIT Press.

Rubenstein, M., A. Cornejo, and R. Nagpal. 2014. "Programmable Self-Assembly in a Thousand-Robot Swarm." *Science*. https://doi.org/10.1126/science.1254295.

Sanan, Siddharth, Peter S. Lynn, and Saul T. Griffith. 2014. "Pneumatic Torsional Actuators for Inflatable Robots." *Journal of Mechanisms and Robotics* 6 (3). https://doi.org/10.1115/1.4026629.

Sasaki, Takao, and Dora Biro. 2017. "Cumulative Culture Can Emerge from Collective Intelligence in Animal Groups." *Nature Communications* 8 (April): 15049.

Schlangen, Erik, and Christopher Joseph. 2009. "Self-Healing Processes in Concrete." *Self-Healing Materials: Fundamentals, Design Strategies, and Applications*. https://onlinelibrary.wiley.com/doi/pdf/10.1002/9783527625376#page=152.

Self-Assembly Lab. n.d. "Liquid Printed Metal." Accessed June 15, 2020. https://selfassemblylab.mit.edu/liquid-printed-metal.

Self-Assembly Lab, Christophe Guberan, Erik Demaine, Carbitex LLC, and Autodesk Inc. n.d. "Programmable Materials." Accessed June 15, 2020. https://selfassemblylab.mit.edu/programmable-materials.

Senatore, Gennaro, Philippe Duffour, Pete Winslow, and Chris Wise. 2017. "Shape Control and Whole-Life Energy Assessment of an 'Infinitely Stiff' Prototype Adaptive Structure." *Smart Materials and Structures* 27 (1): 015022.

Shannon, Claude Elwood. 1940. "A Symbolic Analysis of Relay and Switching Circuits." Massachusetts Institute of Technology. https://dspace.mit.edu/handle/1721.1/11173?show=full?show=full.

Shapiro, Fred R. 1987. "Etymology of the Computer Bug: History and Folklore." *American Speech* 62 (4): 376–78.

Shetterly, Margot Lee. 2018. *Hidden Figures*. HarperCollins.

Sitthi-Amorn, Pitchaya, Javier E. Ramos, Yuwang Wangy, Joyce Kwan, Justin Lan, Wenshou Wang, and Wojciech Matusik. 2015. "Multi-Fab: A Machine Vision Assisted Platform for Multi-Material 3D Printing." *ACM Transactions on Graphics*, no. 129 (July). https://doi.org/10.1145/2766962.

Stanford University. 2015. "Stanford Engineers Develop Computer That Operates on Water Droplets." *Stanford News*. June 8. https://news.stanford.edu/2015/06/08/computer-water-drops-060815/.

Steadman, Ian. 2012. "Brian Eno on Music That Thinks for Itself." *Wired*, September 28. https://www.wired.co.uk/article/brian-eno-peter-chilvers-scape.

"STEM Kits & Robotics for Kids | Inspire STEM Education with Sphero." n.d. Sphero. Accessed June 15, 2020. https://littlebits.com/.

Stephenson, Neal. 1998. *The Diamond Age*. Penguin.

Stoy, Kasper, David Brandt, and David J. Christensen. 2010. *Self-Reconfigurable Robots: An Introduction*. MIT Press.

Sutherland, Ivan E. 1964. "Sketchpad a Man-Machine Graphical Communication System." *Simulation* 2 (5): R–3–R–20.

Tessler, Michael, Mercer R. Brugler, John A. Burns, Nina R. Sinatra, Daniel M. Vogt, Anand Varma, Madelyne Xiao, Robert J. Wood, and David F. Gruber. 2020. "Ultra-Gentle Soft Robotic Fingers Induce Minimal Transcriptomic Response in a Fragile Marine Animal." *Current Biology: CB* 30 (4): R157–58.

Thompson, D'arcy Wentworth. 1917. "On Growth and Form." https://doi.org/10.5962/bhl.title.11332.

Tibbits, Skylar. 2012. "Design to Self-Assembly." *Architectural Design* 82 (2): 68–73.

———. 2017. *Active Matter*. MIT Press.

Tibbits, Skylar J. E. 2010. "Logic Matter: Digital Logic as Heuristics for Physical Self-Guided-Assembly." Massachusetts Institute of Technology. https://dspace.mit.edu/handle/1721.1/64566?show=full.

Tibbits, S., and A. Falvello. 2013. "Biomolecular, Chiral and Irregular Self-Assemblies." http://papers.cumincad.org/cgi-bin/works/Show?acadia13_261.

Tibbits, Skylar, Neil Gershenfeld, Kenny Cheung, Max Lobovsky, Erik Demaine, Jonathan Bachrach, and Jonathan Ward. n.d. "Biased Chains." Accessed June 15, 2020. https://selfassemblylab.mit.edu/biasedchains/.

Turing, Alan Mathison. 1952. "The Chemical Basis of Morphogenesis." *Philosophical Transactions of the Royal Society of London. Series B, Biological Sciences* 237 (641): 37–72.

Vasu, Subeesh, Nimisha Thekke Madam, and Rajagopalan A. N. 2018. "Analyzing Perception-Distortion Tradeoff Using Enhanced Perceptual Super-Resolution Network." http://arxiv.org/abs/1811.00344.

Vella, Dominic, and L. Mahadevan. 2005. "The 'Cheerios Effect.'" *American Journal of Physics* 73 (9): 817–25.

Wehner, Michael, Ryan L. Truby, Daniel J. Fitzgerald, Bobak Mosadegh, George M. Whitesides, Jennifer A. Lewis, and Robert J. Wood. 2016. "An Integrated Design and Fabrication Strategy for Entirely Soft, Autonomous Robots." *Nature* 536 (7617): 451–55.

Wei, Bryan, Mingjie Dai, and Peng Yin. 2012. "Complex Shapes Self-Assembled from Single-Stranded DNA Tiles." *Nature* 485 (7400): 623–26.

Willmann, Jan, Fabio Gramazio, and Matthias Kohler. 2015. "Gramazio Kohler Research, Automated Diversity: New Morphologies of Vertical Urbanism." *Architectural Design*. https://doi.org/10.1002/ad.1989.

Wolfram, Stephen. 2002. *A New Kind of Science*. Vol. 5. Wolfram Media.

Wu, Jiajun, Chengkai Zhang, Tianfan Xue, William T. Freeman, and Joshua B. Tenenbaum. 2016. "Learning a Probabilistic Latent Space of Object Shapes via 3D Generative-Adversarial Modeling." http://arxiv.org/abs/1610.07584.

Wu, Jinrong, Li-Heng Cai, and David A. Weitz. 2017. "Self-Healing Materials: Tough Self-Healing Elastomers by Molecular Enforced Integration of Covalent and Reversible Networks (Adv. Mater. 38/2017)." *Advanced Materials*. https://doi.org/10.1002/adma.201770274.

Yao, Lining, Jifei Ou, Chin-Yi Cheng, Helene Steiner, Wen Wang, Guanyun Wang, and Hiroshi Ishii. 2015. "bioLogic: Natto Cells as Nanoactuators for Shape Changing Interfaces." In *Proceedings of the 33rd Annual ACM Conference on Human Factors in Computing Systems*, 1–10. Association for Computing Machinery.

Zavada, Scott R. Nicholas R. McHardy, Keith L. Gordon, and Timothy F. Scott. 2015. "Rapid, Puncture-Initiated Healing via Oxygen-Mediated Polymerization." *ACS Macro Letters*, July. https://doi.org/10.1021/acsmacrolett.5b00315.

Zhang, Zhizhou, Kahraman G. Demir, and Grace X. Gu. 2019. "Developments in 4D-Printing: A Review on Current Smart Materials, Technologies, and Applications." *International Journal of Smart and Nano Materials* 10 (3): 205–24.

Zykov, Victor, Efstathios Mytilinaios, Bryant Adams, and Hod Lipson. 2005. "Robotics: Self-Reproducing Machines." *Nature* 435 (7039): 163–64.

Index

Page numbers in italics refer to figures.

Image Credits

Page ii: *Fluid-Assembly Chair*
Self-Assembly Lab, MIT

Page vi: *Programmable Textiles*
Self-Assembly Lab, MIT and
Christophe Guberan

Page viii: *Liquid Printed Bag*
Self-Assembly Lab, MIT and
Christophe Guberan

Page xii: *Programmable Wood*
Self-Assembly Lab, MIT,
Christophe Guberan and
Erik Demaine

Page 14: *Logic Matter*
Skylar Tibbits, MIT

Page 40: *Fluid Crystallization*
Self-Assembly Lab, MIT and
Arthur Olson

Page 56: *Slip-Form Rock Jamming*
Self-Assembly Lab, MIT and
Google

Page 76: *The Octobot*
Lori K. Sanders, Harvard
University

Page 96: *Kilobots*
Michael Rubenstein and
Radhika Nagpal, Harvard
University

Page 116: *Programmable Wood*
Self-Assembly Lab, MIT and
Christophe Guberan

Page 134: *Liquid Printed Metal*
Self-Assembly Lab, MIT and
AWTC

Page 154: *Self-Replicating Spheres*
Self-Assembly Lab, MIT

Page 174: *Liquid Printed Light*
Self-Assembly Lab, MIT,
Christophe Guberan and
Marcelo Coelho

Page 180: *Climate-Active Textiles*
Self-Assembly Lab, MIT and
Ministry of Supply.
Photo by Lavender Tessmer

Page 210: *Rock Print*
Gramazio Kohler Research,
ETH Zurich and Self-Assembly
Lab, MIT

Colophon

TYPEFACES
IBM Plex Serif
Whyte Inktrap

TYPESETTING
BookComp, Inc.
Belmont, Michigan

PAPER
120gsm Chen Ming
uncoated

PREPRESS
Jay's Publishers Services, Inc.
Hanover, Massachusetts

PRINTING & BINDING
Asia Pacific Offset
China

EDITORIAL
Jessica Yao
María García

PRODUCTION EDITORIAL
Terri O'Prey

TEXT & COVER DESIGN
Chris Ferrante

PRODUCTION
Steve Sears

PUBLICITY
Sara Henning-Stout
Kate Farquhar-Thomson

COPYEDITOR
Madeleine Adams

Photograph by Olivier Hess

Skylar Tibbits is the founder and codirector of the Self-Assembly Lab and Associate Professor of Design Research in the Department of Architecture at the Massachusetts Institute of Technology. His books include *Active Matter* and *Self-Assembly Lab: Experiments in Programming Matter*. He lives in Boston.

WEBSITE
selfassemblylab.mit.edu

TWITTER
@SkylarTibbits

INSTAGRAM
@skylartibbits